# SYSTEMS NEUROSCIENCE

*ACADEMIC PRESS RAPID MANUSCRIPT REPRODUCTION*

# SYSTEMS NEUROSCIENCE

*edited by*

**Jacqueline Metzler**

***University of Massachusetts***
***Amherst, Massachusetts***

*Center for Systems Neuroscience*
*Executive Committee*
*MICHAEL A. ARBIB*
*WILLIAM L. KILMER*
*D. NICO SPINELLI*

ACADEMIC PRESS New York San Francisco London 1977

A Subsidiary of Harcourt Brace Jovanovich, Publishers

ACADEMIC PRESS, INC.
111 Fifth Avenue, New York, New York 10003

*United Kingdom Edition published by*
ACADEMIC PRESS, INC. (LONDON) LTD.
24/28 Oval Road, London NW1

**Library of Congress Cataloging in Publication Data**

Systems neuroscience.

Includes indexes.
1. Brain–Mathematical models. 2. Nervous system–Mathematical models. 3. Neuropsychology–Mathematical models. 4. System analysis.
I. Metzler, Jacqueline. II. University of Massachusetts at Amherst. Center for Systems Neuroscience.
QP376.S95 591.1'88 77-24174
ISBN 0-12-491850-6

PRINTED IN THE UNITED STATES OF AMERICA

F

# Contents

# Contributors

Numbers in parentheses indicate the pages on which the authors' contributions begin.


SHUN-ICHI AMARI (55, 67, 119, 167), Department of Mathematical Engineering and Instrumentation Physics, University of Tokyo, Tokyo, Japan [CSN Fellow, 1975–1976]

MICHAEL A. ARBIB (119, 221), Department of Computer and Information Science and Center for Systems Neuroscience, University of Massachusetts, Amherst, Massachusetts 01003

ARTHUR I. KARSHMER (55, 197), Department of Computer and Information Science and Center for Systems Neuroscience, University of Massachusetts, Amherst, Massachusetts 01003

WILLIAM L. KILMER (183, 241), Department of Computer and Information Science and Center for Systems Neuroscience, University of Massachusetts, Amherst, Massachusetts 01003

FRED K. LENHERR (197), Center for Systems Neuroscience, University of Massachusetts, Amherst, Massachusetts 01003

ISRAEL LIEBLICH (55, 167, 183, 221), Department of Psychology, The Hebrew University of Jerusalem, Jerusalem, Israel [CSN Fellow, 1975–1976]

JACQUELINE METZLER* (1, 25), Center for Systems Neuroscience, University of Massachusetts, Amherst, Massachusetts 01003 [CSN Fellow, 1975–1976]

THOMAS H. PROBERT (183), Department of Computer and Information Science, University of Massachusetts, Amherst, Massachusetts 01003

D. NICO SPINELLI (25), Department of Computer and Information Science and Center for Systems Neuroscience, University of Massachusetts, Amherst, Massachusetts 01003

DOUGLAS WILLARD (241), Harvard Medical School, Boston, Massachusetts 02115 [CSN Visitor, 1975–1976]

*Present address: Department of Neurosurgery, Yale University School of Medicine, New Haven, Connecticut 06510.

# Preface

Systems Neuroscience is an approach within neuroscience that seeks to bridge the gap between theory and experiment by the judicious development and application of system theory and computer simulation. The Center for Systems Neuroscience was created to further such developments, and this volume records the fruits of the first year (1975–1976) of its activity: a wide range of interdisciplinary contributions to neuroscience. The resident staff at the University of Massachusetts that year were Michael Arbib (Director), William Kilmer, Nico Spinelli, Fred Lenherr (Center Associate), and Arthur Karshmer (Computer Director). They were joined by three Fellows, Shun-ichi Amari, Israel Lieblich, and Jacqueline Metzler, and a Visitor, Douglas Willard, whose diverse backgrounds contributed to the further development of the systems neuroscience methodology.

The papers may be roughly divided into four overlapping categories. The first three articles pertain to vision. Metzler's "Mental Transformations" presents psychophysical data on human visual perception and memory, and indicates the value of a top-down analysis in relating quantitative measurements of human behavior to the fine-level analysis of the neurophysiology. Metzler and Spinelli then analyze the fine details of cortical neurophysiology to provide a neural net model of a mechanism for tilt constancy. Finally, Amari, Lieblich, and Karshmer report their computer analysis of a neural model of masking and flicker fusion.

In the next four papers, the emphasis shifts from computer simulation to mathematical analysis—studying the properties of differential equations that represent networks of neurons. After Amari's survey of his own mathematical approach to neural nets, Amari and Arbib present a detailed analysis of competition and cooperation in neural networks, with applications to sensorimotor transformations in tectum and superior colliculus, to mode selection in reticular formation, and to cortical mechanisms for stereopsis. Leiblich and Amari demonstrate how circuits in the limbic system could explain the kindling phenomenon, while Kilmer, Lieblich, and Probert offer a differential equation that has relevance for ecological as well as neural systems.

The next paper by Karshmer and Lenherr focuses on computer methodology. It is a position paper rather than a finished piece of research, establishing directions for our CORETEX project—the development of an interactive computer graphics language for the simulation of concurrent processes such as those occurring in neural networks. It is hoped that this paper will generate feedback from the neuroscience community regarding appropriate goals for computer simulation methodology.

The final section seeks to understand how the brain represents the spatial dimensions of the world around it. Continuing the approach adopted in the first paper of this volume, Arbib and Lieblich develop appropriate data structures for internal representations that accord well with observations on the motivated learning of spatial behavior by rats. Kilmer and Willard, on the other hand, contribute a framework for the analysis of recent experiments on the possible embodiment of cognitive maps within the hippocampus and its environs.

We believe that these papers attest to the validity of the interdisciplinary approach. Nonetheless, we feel that much further development of systems neuroscience is required. The nascent stage of the CORETEX project is typical of the field and suggests how much ad hoc technique there is in computer simulation today. Tools to determine the most efficient simulation of a given network are lacking—mere iteration of the network equations is surely not always the most efficient way to get the computer to answer our questions about the network. The field is still at a stage when many models of a given neural system are developed by different researchers without there being any consistent methodology for comparing their relative merits. This is compounded by multiple problems: not enough modelers suggest critical experiments that can be conducted in the laboratory; not enough experimentalists are motivated to explore the implications of extant models; and, even when data exist on the strengths and weaknesses of several models, researchers often lack the methodology to synthesize the strengths of these models into a more powerful and elegant one. Finally, there remains in systems neuroscience a continuing need for new concepts, new computer simulation tools, and new techniques of mathematical analysis. Even more important, we must bring these to bear in the juxtaposition of theory and experiment, with the outcome being the development of models that broaden our understanding of the brain. This way lies the healthy cumulative development of systems neuroscience.

The activities of this first year were funded by a grant from the Alfred P. Sloan Foundation. Their generous support is gratefully acknowledged. Our special thanks go to Linda M. Strzegowski for her competent and dedicated assistance in preparing the manuscript.

JACQUELINE METZLER
MICHAEL A. ARBIB
*Amherst, Massachusetts*
*March 1977*

# Mental Transformations: A Top-Down Analysis

JACQUELINE METZLER*
*Center for Systems Neuroscience*
*University of Massachusets at Amherst*

ABSTRACT *Evidence that the visual system can model the physical environment--i.e., that it exhibits processes that have a similar relational structure to physical space and are largely analog in nature--is found in a series of recent experiments by Shepard and Metzler and their colleagues. This paper provides a top-down analysis of the results of these studies of mental transformations in an attempt to bridge the gap between pyschological studies of learning, memory, and other cognitive processes, and neurophysiological investigations of the properties of single neurons. In Section 1, the ability to recognize rotated objects--or constancy of perceived shape under rotation--is compared with the traditional constancies of color, size, and location. Section 2 considers the possible strategies for recognizing rotated forms and provides empirical support for a theory of mental rotation. The neural substrates of pattern recognition and higher level processes are discussed in Section 3; and in Section 4, the visual system--and perceptual systems, in general--are considered as generative devices which, in addition to detecting perceptual invariances, can construct certain types of perceptual experiences.*

## *1. INTRODUCTION*

Our visual world remains relatively constant despite the large scale changes in the energy patterns that reach our retinas.

---

*1975-76 Fellow, presently at the Department of Neurosurgery, Yale University School of Medicine, New Haven, CT.

In other words, some things in our environment maintain certain *invariant properties* with respect to changes in physical energy. These perceived invariant properties are referred to as the "perceptual constancies." Traditionally, these invariant phenomena included brightness and color, size, shape, and translation (i.e., location) constancy. They were all grouped under the rubric "constancy" because in each situation the perceived or psychological dimension does not change--i.e., it remains constant--following changes in the physical dimension.

The ability to recognize an object as the same object despite changes in its orientation can also be considered as a kind of perceptual constancy, namely, constancy of perceived shape following rotation in space. Like the traditional and better understood constancies of color, size, and location, the perceived dimensions (orientation) of an object, when viewed from a given position, do not correspond to the planar projection (orientation) of the object on the retina. For example, an "ᴚ" is perceived as an "R", and  and  are both perceived as a cube.* Changes in the orientation of an object that result from rotations about an axis coinciding with the line of sight (i.e., an axis through the point of projection) are like most other rigid spatial transformations (size and location) in that there is an invariant or direct relationship between the changes in the three-dimensional object and the corresponding change in the retinal image. However, for changes in orientation that result from rotations in depth (i.e., about any axis other than the axis through the point of projection), the relationship between the change in the object and the change in the retinal image is less direct. Although the three-dimensional object itself remains rigid under rotation, the two-dimensional projection undergoes a relatively complex, nonrigid

---

*While it is true that a ◇ is perceived as being different from a □ and a "6" is not seen as a "9", this is the result of their use as symbols whose meanings depend on their *orientations* as well as their shapes.

deformation. Parts of the object as projected may shrink or deform relative to other parts, or they may even disappear completely, only to reappear on the other side of the two-dimensional projection.

In addition, objects that differ either in brightness or color can generally be recognized as the same object as quickly as if they were of identical brightness or color, while those that differ either in size or location are often recognized as equivalent *almost* as quickly as if they were identical in size or location, but the recognition process does require some time (Sekuler and Nash, 1972; Metzler, unpublished results). However, object equivalence is more difficult to determine--requiring additional time and, presumably, more "mental" effort--when the two objects require rotation in the two-dimensional plane to bring them into congruence (Cooper and Shepard, 1973a,b; Cooper, 1975), and the task becomes increasingly more difficult when the objects differ by rotations in depth (Shepard and Metzler, 1971; Metzler and Shepard, 1971, 1974; Metzler, 1973). The difficulty of the task may be explained in part by the following observations. Brightness and color constancy are based primarily on the retinal distribution of light energy, and the role of learning is small. In contrast, size, translation, and shape constancy are also influenced by experience, attitude, and therefore, presumably, the mediational role of the cortex. Size and shape constancies show improvement with learning although size constancy demonstrates more rapid and complete learning than shape constancy. Thus, we can order these three constancies from primitive to complex--brightness being the most primitive, size in the middle, and shape the most complex. It is interesting to note that the same ordering is found if we use phylogenetic development as a criterion (see, e.g., Köhler, 1915; Brunswik, 1929; Fields, 1932; Gellerman, 1933; Lashley, 1938).

The fact that judgments of equivalence require time and mental effort when the objects differ in size and, especially, in orientation and shape implies that the visual pathways are not merely passive conduits that transmit available stimulation from the retina to higher centers in the brain. Instead, the visual system--and other perceptual systems, as well, as we shall note in Section 4--are flexible, generative devices which, in addition to detecting invariants in changing stimulation, construct certain perceptual experiences.

Shepard and Metzler (1971) reasoned that something of interest could be learned about how objects--in particular, three-dimensional objects--are internally represented by observing the way in which subjects deal with such objects presented in different orientations. We were especially interested in determining whether the internal representation is structurally more similar to the two-dimensional projection of the object or whether it is more closely related to the three-dimensional object itself. In the first of several experiments designed to measure the time required by human subjects to respond discriminatively to spatially transformed visual objects, Shepard and Metzler (1971) presented subjects with pairs of perspective line drawings of three-dimensional objects. Each object consisted of ten solid cubes attached face-to-face to form a rigid asymmetrical structure with two free ends and three right-angled bends. The subject's task was to determine as quickly as possible whether the two objects portrayed were of the same or of different three-dimensional shape. We found the time required to determine that the two drawings represented objects of the same three-dimensional shape (i) increased linearly with the angular difference in the portrayed orientation of the two objects and (ii) was relatively independent of whether the angular difference was produced by a rigid rotation of one of the two-dimensional drawings in its own picture plane or whether it was produced by the two-dimensionally, much more complex, nonrigid transformation corresponding to a rigid rotation of the three-dimensional object in depth.

## 2. *RECOGNITION OF ROTATED OBJECTS*

### 2.1 *Possible Strategies for Determining the Identity of Shape of Rotated Objects*

The first of the two findings reported by Shepard and Metzler suggests that the internal process proceeds at a *constant rate*. The second result indicates that this process proceeds at the *same rate*, independent of whether the angular difference corresponds to a rotation in the picture plane or in depth. Several possible theories or strategies might be offered in way of explaining these results. The simplest explanation that appears to account for both findings is that the subject constructs an internal representation of the three-dimensional object on the basis of the two-dimensional drawing and then carries out on this internal model, at a more or less constant and bounded rate, an analog of a physical rotation of the corresponding external object in order to test for the congruence of the pair. Moreover, the results suggest that subjects carry out the mental operations upon internal representations that are more analagous to the three-dimensional objects depicted by the line drawings than to the two-dimensional drawings actually presented.

Some of the alternative, nonrotational theories that might be proposed to explain how subjects determine the identity of rotated three-dimensional shapes include:

i. One in which the subject generates a rotationally invariant structural code for each of the two objects independently and then compares these codes for a match or mismatch;

ii. One in which the subject compares the two objects, feature by feature, ultimately determining whether or not all of the corresponding features of the two objects result in a suitable match;

iii. An operation requiring the subject to solve a series of equations, for example, before determining the congruence or noncongruence of the pair; or, perhaps,

iv. One requiring a search process that becomes progressively more difficult as the difference in portrayed orientation increases.

It seems, however, that all of these alternative theories are rendered implausible by the results. For example, the first alternative, which suggests that the two objects are considered separately at first and only later compared for a match or mismatch, would predict reaction times to be independent of angular departure; i.e., the reaction-time function would be essentially horizontal rather than linearly increasing. And while any of the alternative theories might imply that the increase in reaction time simply reflects the increasing difficulty in processing the stimuli as the two views become more and more disparate owing to changes in portrayed orientation, it appears that these theories cannot explain either (i) why the increasing reaction time functions are so precisely linear, or (ii) why the intercepts and slopes of the functions for rotations in depth and in the picture plane are so very similar. [For a detailed account of the theory of mental rotation, see Metzler and Shepard (1974) and Shepard (1975). Some of the alternative, nonrotational theories mentioned above are discussed more fully in Metzler and Shepard (1974).]

A theoretical distinction can also be drawn between the various theories that might be offered in order to explain Shepard and Metzler's findings. The theory that accounts very nicely for both results, namely, that of mental rotation, implies an *analog process*. That is to say, in the course of such a process, the internal representation passes through a certain trajectory of intermediate stages, each of which bears a one-to-one correspondence to an intermediate stage of an external physical rotation of the object. The nonrotational theories, on the other hand, imply a process that is not in any useful sense analogous to the external state of affairs.

### 2.2 *Additional Empirical Support for a Theory of Mental Rotation*

In subsequent experiments by Shepard and Metzler and their colleagues, significant progress has been made in demonstrating that internal representations can be mentally transformed in ways analogous to the kinds of transformations that can be performed on the corresponding external objects. Several additional findings emerged from two subsequent experiments on mental rotation of three-dimensional objects by Metzler (1973). In these experiments, subjects were again presented with perspective views of differently oriented three-dimensional structures. The objects, however, were simplified forms of those used in the initial study, consisting of only seven cubes and only two right-angled bends. In addition, for pairs differing by rotation in depth, the rotation was always about the natural axis of the object (i.e., about an axis through the central and longest straight row of cubical blocks).

The first experiment, which was similar in basic procedure to the initial study except that certain deliberate manipulations were used to induce subjects to rotate the long way around (i.e., beyond $180^{\circ}$) on some trials, yielded two significant results. The first concerned the reaction times for those pairs in which the two objects were isomeric rather than identical. Although the angle through which differently shaped three-dimensional objects must be rotated to achieve congruence cannot be defined, the following assumption was made (owing to the manner in which the stimuli were constructed): The subject carried out a mental rotation in order to bring one part of the object into congruence with the corresponding part of the other object and then checked to see whether the other part(s) of the objects--namely, one cube, in this case--had also been brought into congruence by this operation. Reaction times for the "different" pairs were plotted as a function of the rotation (of the longer arm of the object) required to achieve *partial* congruence. While the decision latencies for the "different" pairs were about 200 to 300 msec longer than those for the "same" pairs, the reaction time functions were very similar in both cases. Reaction time was again found to increase linearly with the angular difference in the orientation of the longer arms of the two objects.

The second new finding concerned those pairs of objects in which the direction of shortest rotation to bring the two objects into congruence was unexpectedly reversed so that if the subjects continued to rotate in the designated direction, they would eventually rotate beyond $180^{\circ}$ or the long way around. Metzler found that reaction times to pairs differing by $225^{\circ}$* in the designated direction were bimodally distributed with an upper mode corresponding to the linearly predicted value for rotations of $225^{\circ}$ and a lower mode centered at the time required to rotate $135^{\circ}$. Individual subjects differed considerably in their tendency to go the long or short way around.

In the second experiment (Metzler, 1973), the presentation of the second perspective view was delayed until the required mental rotation of the object depicted in the first perspective view could be carried out. Since the required rotation would be essentially completed, the subject should then be able to make a direct template-like match between the transformed internal representation of the first stimulus and the externally presented second stimulus. Consequently, the reaction time, when measured from the onset of the second stimulus, should be uniformly short and independent of the angular difference between the two objects. These expectations, based on the theory of mental rotation, were essentially confirmed.

The results of these experiments by Metzler also revealed that reaction times were shorter for rotations about an axis that coincided with the natural axis of the object itself. There was

*Pairs differing by more than $180^{\circ}$ were confined to angular departures of $225^{\circ}$ in the designated direction, which was the next step beyond $180^{\circ}$ owing to the $45^{\circ}$ increments used in this study.

also a slight but consistent tendency towards shorter reaction times for rotations about a vertical axis, with rotations about inclined and horizontally projected axes requiring longer times, respectively [cf. results of experiments by Shepard and Feng (see Shepard, 1975) and Taylor (1972) involving imagined reflections of two-dimensional objects about vertical and horizontal axes].

In an experiment by Shepard and Feng (1972) involving sequences of spatial operations about different axes, subjects were presented a picture of one of the patterns of six connected squares produced when the faces of a cube are unfolded and laid out flat. Subjects were to determine, as quickly as possible, whether two marked edges of two different squares would meet if the squares were folded back into a cube. Decision time was found to increase linearly with the sum of the number of squares that would have been involved if the operation of folding up the squares had actually been carried out.

Related experiments have focused on transformed two-dimensional alpha-numeric characters and random and familiar shapes. The results of these investigations have revealed that (i) the time required to determine that a spatially transformed object is of the same shape as a comparison object is a monotonically (Cooper and Shepard, 1973a,b, 1975) and often linearly (Cooper, 1975) increasing function of the angular difference between the two shapes; (ii) when the subject is prepared to mentally rotate the internal model in a particular direction, reaction time continues to increase linearly beyond $180^{o}$ (Cooper, 1975); (iii) if the subject receives advance information regarding both the object to be presented and the orientation in which it will appear, reaction time becomes uniformly short and independent of orientation (Cooper and Shepard, 1973a, 1975; Cooper, 1975); (iv) the time required to prepare for a rotated object is essentially the same as that required to respond to such a rotated stimulus in the absence of advance preparation (Cooper and Shepard, 1973b; Cooper, 1975); and (v) information concerning the orientation in which the object is to appear is insufficient to allow subjects to prepare for a rotated object; advance information concerning the object's identity must also be provided (Cooper and Shepard, 1973b). The results of these experiments corroborate the findings from the studies on the rotation of three-dimensional objects.

Although one certainly cannot rule out the possibility that a nonrotational theory will be proposed which will be capable of explaining all of the empirical findings reviewed in the preceding paragraphs, for the present, the theory of mental rotation appears to provide the simplest and most complete account of the principal results. (For a detailed discussion of the theoretical implications of these experimental findings, see Shepard, 1975.) On the basis of the evidence provided by these and other studies (see, e.g., Huttenlocher, 1968; Shepard and Chipman, 1970; Kosslyn, 1973) of spatial images, I would like to propose that the neural spatial representation afforded by the visual system in which these internal processes occur is a model or an analog of physical space.

### 3. *NEURAL SPATIAL REPRESENTATION*

Perhaps one of the most challenging pursuits in studying the visual system is to try to discover the mechanisms of perception and thus relate data from psychological studies of higher level cognitive processes to the results of neurophysiological investigations of the properties of single neurons. However, before considering what might be a plausible neural mechanism for transforming internal representations of spatial stimuli, let us first consider the components of the system, i.e., the hardware of the machine--the individual neurons. In general, the cellular analogs of learning, memory, and other cognitive processes, such as visual imagery, are difficult to determine and assess, since the integrated and purposeful behavior of any animal with experience is extremely complicated. Nevertheless, the results of several recent experiments designed to modify the functional properties of cortical cells through selective visual experience suggest that it may be possible to obtain direct expression of the internal representation of visual patterns by systematically mapping the receptive fields of single units along the visual pathways.

The first of the many studies of the influence of early visual experience on the organization of cortical neurons in the developing brain was that of Hirsch and Spinelli (1970, 1971). Kittens were raised with one eye viewing vertical bars and the other eye viewing horizontal bars. Following the rearing period, all cortical neurons with elongated receptive fields were found to respond to input from only one eye. In addition, the receptive field orientations were coincident with the bar pattern to which the eye had been exposed. A short time later Blakemore and Cooper (1970) reported a similar finding. They limited the binocular environment of kittens to only vertical or horizontal stripes. After the exposure period, they found that when stimuli were presented in the orientation of the earlier visual experience, cortical neurons produced essentially normal neurophysiological responses. However, no responsive neurons were encountered when stimuli were oriented perpendicular to the orientation the kittens had experienced.* The results of these and other studies (e.g., Shlaer, 1971; Spinelli and Hirsch, 1971; Spinelli, Hirsch, Phelps, and Metzler, 1972; Cynader, Berman, and Hein, 1973; Pettigrew and Freeman, 1973; Spinelli, Metzler, and Phelps, 1973, 1975; Metzler and Spinelli, 1975, in preparation; Schechter and Murphy, 1976) that have selectively controlled the visual experience of young organisms demonstrate quite clearly that functional neural connections can be modified in a predictable and selective manner by the visual enviroment.

---

*While the results of those studies limiting the early visual environment of kittens to stripes of a single orientation have recently been questioned by Stryker and Sherk (1975), these same investigators have confirmed the findings of Hirsch and Spinelli.

The kittens in Hirsch and Spinelli's initial study were revived and allowed to view the normal environment with both eyes for up to 19 months. The results of the rerecordings at the end of this time (Spinelli, Hirsch, Phelps, and Metzler, 1972) revealed a substantial increase in the percentage of those classes of cells that were either weak or absent following the period of selective visual experience. For example, many disc-shaped receptive fields were now encountered. Moreover, these receptive fields, acquired during binocular viewing, were generally binocular. The data, however, also revealed that cells whose response properties mimicked the stimuli viewed during development were almost completely unaffected by the subsequent visual experience. These cells were monocularly driven and had receptive field orientations corresponding to the orientation of the lines to which the eye activating the unit had been exposed during rearing. These results demonstrate that, for at least some cells in visual cortex, the modifications generated by visual experience remain as permanently acquired characteristics. It appears, therefore, that once a mechanism is established, it persists.

The results of these studies of the effects of the early visual environment have far-reaching implications for the development of the human visual system. Indeed, a number of recent experiments with humans (e.g., Freeman, Mitchell, and Millodot, 1972; Annis and Frost, 1973; Freeman and Thibos, 1973; Mitchell, Freeman, Millodot, and Haegerstrom, 1973; Banks, Aslin, and Letson, 1975; Freeman, 1975) provide evidence that visual acuity, accommodation, binocularity, and even orientation specificities may be affected by early experience.

In a recently completed study (Metzler and Spinelli, in preparation; see, also, Metzler and Spinelli, this volume), we have successfully isolated a tilt constancy mechanism in the kitten visual cortex. A number of investigators (Horn and Hill, 1969; Spinelli, 1970a; Denney and Adorjani, 1972; Horn, Strechler, and Hill, 1972) have reported that a small number (about six percent) of the neurons in the visual cortex of the cat seem to change their orientation selectivity during head or body tilt so as to compensate for the change in head or body position. In an effort to examine the effects of experience on the development of tilt constancy, kittens were reared with one eye viewing two vertical bars, while the other eye viewed two horizontal bars. The two bars presented to one eye were fixed in the goggles each kitten wore so that they always maintained a constant orientation on the retina, even when the kitten's head was tilted. The two bars viewed by the other eye were gravity-stabilized so that their orientation remained constant, moving over the retina whenever the kitten rotated its head.

Following the exposure period, we found that all cortical (Area 17) neurons with elongated receptive fields were monocularly activated, and when mapped with the kitten in the standard (horizontal) position, the receptive field orientation corresponded to the orientation of the bars to which the eye had been exposed. When the kitten was rotated about its longitudinal axis, the receptive field orientation of cells activated by the eye that had viewed the pattern having a fixed retinal orientation during development rotated by the same amount and in the same direction; i.e., these receptive fields were not tilt constant. The orientation of those receptive fields that could be mapped only through the eye that had been exposed to the gravity-stabilized pattern during development, on the other hand, remained unchanged following body tilt; i.e., these receptive fields appeared to demonstrate tilt constancy.

The data from this experiment, together with the results of other studies of selective visual experience, were used in designing a neural net model of cortical development and tilt constancy (see Metzler and Spinelli, this volume). However, before higher-level transformations, such as mental rotation, can be modeled, or even studied neurophysiologically, we will want to give more thought to the formation and transformation of more complex internal representations and to the relationship between the higher- and lower-level processes. The hope is that models of adaptive neural processes and low-level transformations may provide insight into the design of appropriate neural networks and critical experiments which may ultimately result in the neurophysiological explication of higher cognitive functions, including mental rotation.

As we have noted, most of the neurons found in kittens reared under controlled conditions develop receptive field properties that appear to be a direct consequence of the experience the animal encountered early in its life. We believe these receptive field characteristics--shape and orientation, in particular--are generated by neurons belonging to memory networks (see Spinelli, Hirsch, Phelps, and Metzler, 1972). These memory networks may be analogous to the neuronal local circuits as described by Rakic (1975), and Schmitt, Dev, and Smith (1976). It seems that the small neurons which give rise to most of the dendritic processes comprising the local circuits develop after the genetically determined, long-axon projection neurons are formed. Jacobson (1975) argues that local circuit neurons thereby possess a greater degree of modifiability. Furthermore, neural local circuits appear to be regionally differentiated with respect to such plastic properties.

Schmitt et al. suggest that the highly intricate aggregates of local circuits "may indeed prove to be the neural substrates of higher brain function" (p. 119), a possibility not inconsistent with our interpretations (see, also, Cajal, 1937). Although neuronal local circuits cannot at the present time elucidate the nature of cognitive events, they do possess a number of properties that would appear to be requisite features of any system capable of mediating higher-level processes. For example, local circuits are characterized by a high density of computational structures, the potential for complex interactions, and the rapid processing of information, owing to the tight interweaving of fine neuronal processes and the short intra- and inter-neural distances over which they operate. In addition, their use of low-level graded potentials requires high sensitivity, permitting discriminations that cannot be equaled by an all-or-none process--and endows them with the capability of explaining the analog nature of mental transformations.

To speak of the analog nature of mental transformations is to imply that the neural activity occurring when one compares two differently oriented objects is very similar to the neural process taking place when one actually observes one external object physically rotating into congruence with a second object. It implies further that in the course of such a process, the internal representation passes through a trajectory of intermediate states, each of which bears a one-to-one correspondence to an intermediate stage of the rotation of the external object. It seems that this can be achieved only by carrying out a series of small adjustments in orientation, each of which preserves the essential structure of the object until the desired orientation is reached.

Such small adjustments in orientation might be achieved by the successive activation of a number of local circuits, each of which is driven by a number of the same neurons as its temporal neighbors, but at the same time is receiving input from a slightly different subset of neurons than the circuits activated either before or after it. A computer analog of this process would be a number of *forwardly linked lists*, i.e., memory traces, with each *list* representing the object in a different orientation, and the *links* representing the next position in an ordered succession of orientations representing the rotation being executed. In this way, different rotations could be achieved simply by changing the links, while the *number* of lists to be accessed would vary directly with the size of the angular discrepancy. The time required by the process would, therefore, be a linearly increasing function of the angular difference in orientation. An alternative would be to have a *single* list or memory trace which is then "rotated" by a *3-D* graphics program. However, given the difficulty in writing programs and the abundance of memory circuits in the biological system, it appears that the first hypothesis may be the more likely one.

## 4. *PERCEPTUAL SYSTEMS AS GENERATIVE SYSTEMS*

The "perceptual systems" as defined by Gibson (1966) are active systems that register the invariant properties of stimulation available to the receptors and provide the observer with direct knowledge of his environment. In other words, Gibson has substituted perceptual systems for the senses, and his theory has had far-reaching implications for a general theory of perception. The idea I would like to consider here is that perceptual systems are adaptable apparatuses which, in addition to detecting invariants in changing stimulation, can construct certain types of perceptual experience.

Certainly there is nothing novel in thinking of the perceptual systems as flexible, adaptable, generative devices. This viewpoint also expresses the position advanced by such constructivists as Neisser (1967), Kolers (1968), and Gregory (1972). According to the constructivist view, perceptual experience is not a direct response to stimulation. It is, instead, created from several components, only some of which are provided by the input to the receptors. Other factors influencing perceptual experience include, for example, our motives, expectations, biases, etc. Constructivism in its extreme form considers all perceptual experiences to be constructed from bits and pieces drawn from memory, which are themselves constructed from fragments of past experiences (Gregory, 1972).

The theory advanced by the constructivists, and particularly that put forth by the extremists among them, fails to appreciate the richness of available stimulation at the receptor level as well as the intricacies of the perceptual machinery itself. On the other hand, Gibson's principle of direct perception denies our predictions of, or expectations about, perceptual events which are derived from our internal model and knowledge of the world. An apt description of the perceptual systems appears to lie somewhere between these two views; i.e., we may consider as the primary function of the perceptual apparatus, that of detecting the invariant structures in available stimulation, while at the same time recognizing the fact that the systems may create experiences in the absence of such stimulation. Imagining, dreaming, and hallucinating are most certainly examples of this latter characteristic of the perceptual machinery.

This raises the question as to whether the mechanisms employed in generating experiences in the absence of stimulation are indeed the same as, or in any way overlap with, the normal mechanisms of perception. Segal and Fusella (1970, 1971) instructed subjects to imagine either a visual or auditory event and to indicate when they had formed a good image. At that point, a signal was either presented or not, and the subjects' task was to report whether or not the signal occurred. Segal and Fusella found that the detection of a visual signal was impaired more significantly by concurrent visual imagery than by concurrent auditory imagery, and vice versa. The inference in this case is that detecting a visual signal and generating a visual image are functions performed by a common (or, at least, overlapping) mechanism(s). Similar conclusions can be drawn from the results of other investigations (see, e.g., Brooks, 1967, 1968; Atwood, 1971). We can infer, on the basis of this evidence, that detecting or perceiving a sensory event and imaging the same involve, at least at some level, the same neural structures.

Related to the generative potential of the visual system are the experiments conducted by Shepard and Metzler and their colleagues. In these experiments, the subjects model covertly the operations that would be performed if they were actually perceiving the external event. Moreover, the internal events are, as we noted earlier, analogs of the external events for they are carried out in continuous space and real time. It is on this evidence that we have argued that the neural spatial representation of our visual environment is a model or an analog of the physical world.

There is also neurophysiological data to support the idea that a perceptual system can be generative. John and Killam (1959) used flickering light as the conditioned stimulus for establishing a conditioned avoidance response. Records from visual cortex of a cat trained to press a bar to avoid a shock when a 4-cps flickering light was presented revealed 4-cps activity. If a 10 per second flicker was then presented, the animal still performed the avoidance response, but the cortical records appeared to be a mixture of 10-cps and 4-cps activity. Following differentiation to the 10 per second flicker, cortical activity was largely represented by 10 cps. Thus it appears that, while the cat was generalizing to the 10 cps, two kinds of activity occurred in the visual cortex; namely, the 10 cps produced by the stimulus and the 4 cps, conceivably a reconstruction of what the animal had learned previously. The inference is that the cat performed appropriately to the 10 cps because the previous (4-cps) memory was being played back with the appropriate behavior attached.

Waveforms signalling the presence or absence of reinforcement and the type of behavior to be performed have also been observed by Pribram, Spinelli, and Kamback (1967). Similarly, Morrell (1961), in his experiments on the mirror epileptic focus, reported that, after several weeks of stimulation at a certain frequency, activity in the mirror point persisted even after the stimulation was discontinued. Again, the implication is that neural structures have the capacity to generate the temporal pattern of previous stimulation.

A neural model of such a generative system in the central nervous system is Spinelli's (1970b) OCCAM (Omnium-gatherum Core Content-Addressable Memory). In this model, memory networks are addressed in parallel by stimuli entering the CNS. The networks form content-addressable memories, wherein providing the system with part of the information will enable it to play back the entire sequence. The tilt-constancy model developed by Metzler and Spinelli (this volume) incorporates many of the same features that characterize OCCAM. Parsimonious models that strive to remain as faithful as possible to the neurophysiological data have been found to be extremely helpful in generating hypotheses which can be tested by further experimentation. An answer to the question: "What is the structure of the neuronal network that can perform mental transformations?" may be provided by the specific assumptions resulting from an extension of such adaptive neural systems as OCCAM and Metzler and Spinelli's tilt-constancy model. Such an extension might, for example, incorporate the properties of local circuit neurons in establishing the neural substrate for higher level cognitive events.

ACKNOWLEDGMENTS *This paper was prepared during the author's tenure as a Fellow at the Center for Systems Neuroscience, which is funded by a grant from the Alfred P. Sloan Foundation. The author would like to thank Michael Arbib, William Kilmer, Philip Marks, and D. Nico Spinelli for their helpful suggestions concerning the manuscript, and Richard Roemer for his valuable comments on an earlier draft of this work.*

*REFERENCES*

Annis, R. C., and Frost, B. 1973. Human visual ecology and orientation anisotropies in acuity. *Science 182:* 729-731.

Atwood, G. 1971. An experimental study of visual imagination and memory. *Cognitive Psychology 2:* 290-299.

Banks, M. S., Aslin, R. N., and Letson, R. D. 1975. Sensitive period for the development of human binocular vision. *Science 190:* 675-677.

Blakemore, C., and Cooper, G. F. 1970. Development of the brain depends on the visual environment. *Nature 228:* 477-478.

Brooks, L. R. 1967. The suppression of visualization by reading. *Quarterly Journal of Experimental Psychology 19:* 289-299.

Brooks, L. R. 1968. Spatial and verbal components of the act of recall. *Canadian Journal of Psychology 22:* 349-368.

Brunswik, E. 1929. Zur Entwicklung der Albedowahrnehmung. *Zeitschrift für Psychologie 109:* 40-115.

Cajal, S. Ramón y. 1937. *Recollections of My Life*. Translated by E. Horne Craigie. Philadelphia: The American Philosophical Society.

Cooper, L. A. 1975. Mental transformation of random two-dimensional shapes. *Cognitive Psychology 7:* 20-43.

Cooper, L. A., and Shepard, R. N. 1973a. The time required to prepare for a rotated stimulus. *Memory and Cognition 1:* 246-250.

Cooper, L. A., and Shepard, R. N. 1973b. Chronometric studies of the rotation of mental images. In W. G. Chase, ed., *Visual Information Processing*. New York: Academic Press.

Cooper, L. A., and Shepard, R. N. 1975. Mental transformations in the identification of left and right hands. *Journal of Experimental Psychology: Human Perception and Performance 104:* 48-56.

Cynader, M., Berman, N., and Hein, A. 1973. Cats reared in stroboscopic illumination: Effects on receptive fields in visual cortex. *Proceedings of the National Academy of Sciences 70:* 1353-1354.

Denney, D., and Adorjani, C. 1972. Orientation specificity of visual cortical neurons after head tilt. *Experimental Brain Research 14:* 312-317.

Fields, P. E. 1932. Studies in concept formation. I. The development of the concept of triangularity by the white rat. *Comparative Psychology Monograph 9:* 1-70.

Freeman, R. D. 1975. Asymmetries in human accommodation and visual experience. *Vision Research 15:* 483-492.

Freeman, R. D., Mitchell, D. E., and Millodot, M. 1972. A neural effect of partial visual deprivation in humans. *Science 175:* 1384-1386.

Freeman, R. D., and Thibos, L. 1973. Electrophysiological evidence that early abnormal visual experience can modify the human brain. *Science 180: 876-878.*

Gellerman, L. W. 1933. Form discrimination in chimpanzees and two-year-old children. I. Form (triangularity) per se. *Journal of Genetic Psychology 42:* 3-27.

Gibson, J. J. 1966. *The Senses Considered as Perceptual Systems.* Boston: Houghton Mifflin.

Gregory, R. 1972. Seeing as thinking: An active theory of perception. *London Times Literary Supplement,* June 23, pp. 707-708.

Hirsch, H. V. B., and Spinelli, D. N. 1970. Visual experience modifies distribution of horizontally and vertically oriented receptive fields in cats. *Science 168:* 869-871.

Hirsch, H. V. B., and Spinelli, D. N. 1971. Modification of the distribution of receptive field orientation in cats by selective visual exposure during development. *Experimental Brain Research 13:* 509-527.

Horn, G., and Hill, R. M. 1969. Modifications of receptive fields of cells in the visual cortex occurring spontaneously and associated with bodily tilt. *Nature 221:* 186-188.

Horn, G., Strechler, G., and Hill, R. M. 1972. Receptive fields of units in the visual cortex of the cat in the presence and absence of bodily tilt. *Experimental Brain Research 15:* 113-132.

Huttenlocher, J. 1968. Constructing spatial images: A strategy in reasoning. *Psychological Review 75:* 550-560.

Jacobson, M. 1975. Development and evolution of type II neurons; conjectures a century after Golgi. In M. Santini, ed., *Golgi Centennial Symposium.* New York: Raven Press.

John, E. R., and Killam, K. F. 1959. Electrophysiological correlates of avoidance conditioning in the cat. *Journal of Pharmacology and Experimental Therapy 125:* 252-274.

Köhler, W. 1915. Optische Untersuchungen am Schimpansen und am Hahnshühn. *Abhandlungen der Königliche-Preussische Akademie der Wissenschaften.* Physikalisch-mathematische klasse, No. 3.

Kolers, P. A. 1968. Some psychological aspects of pattern recognition. In P. A. Kolers and M. Eden, eds., *Recognizing Patterns.* Cambridge: MIT Press.

Kosslyn, S. M. 1973. Scanning visual images: Some structural implications. *Perception and Psychophysics 14:* 90-94.

Lashley, K. S. 1938. The mechanism of vision. XV. Preliminary studies on the rat's capacity for detail vision. *Journal of General Psychology 18: 123-193.*

Metzler, J. 1973. Cognitive analogues of the rotation of three-dimensional objects. Unpublished doctoral dissertation, Stanford University.

Metzler, J., and Shepard, R. N. 1971. Mental correlates of the rotation of three-dimensional objects. Paper presented at the annual meeting of the Western Psychological Association. San Francisco, April 1971.

Metzler, J., and Shepard, R. N. 1974. Transformational studies of the internal representations of three-dimensional objects. In R. L. Solso, ed., *Theories of Cognitive Psychology: The Loyola Symposium*. Potomac, Maryland: Lawrence Erlbaum Associates.

Metzler, J., and Spinelli, D. N. 1975. Physiological correlates of early visual experience. Paper presented at the annual meeting of the Society for Neuroscience. New York, November 1975.

Metzler, J., and Spinelli, D. N. 1977. Tilt constancy mechanisms in the developing visual cortex (in preparation).

Mitchell, D. E., Freeman, R. D., Millodot, M., and Haegerstrom, G. 1973. Meridional amblyopia: Evidence for modification of the human visual system by early visual experience. *Vision Research 13: 535-558.*

Morrell, F. 1961. Lasting changes in synaptic organization produced by continuous neuronal bombardment. In J. F. Delafresnaye, A. Fessard, and J. Konorski, eds., *Symposium on Brain Mechanisms and Learning*. Oxford: Blackwell.

Neisser, U. 1967. *Cognitive Psychology*. New York: Appleton-Century-Crofts.

Pettigrew, J. D., and Freeman, R. D. 1973. Visual experience without lines: Effect on developing cortical neurons. *Science 182: 599-601.*

Pribram, K. H., Spinelli, D. N., and Kamback, M. C. 1967. Electrocortical correlates of stimulus response and reinforcement. *Science 157: 94-96.*

Rakic, P. 1975. Local circuit neurons. *Neurosciences Research Program Bulletin 13(3): 291-446.*

Schechter, P. B., and Murphy, E. H. 1976. Brief monocular visual experience and kitten cortical binocularity. *Brain Research 109:* 165-168.

Schmitt, F. O., Dev, P., and Smith, B. H. 1976. Electrotonic processing of information by brain cells. *Science 193:* 114-120.

Segal, S. J., and Fusella, V. 1970. Influence of imaged pictures and sounds on detection of auditory and visual signals. *Journal of Experimental Psychology 83:* 458-464.

Segal, S. J., and Fusella, V. 1971. Effect of images in six sense modalities on detection of visual signal from noise. *Psychonomic Science 24:* 55-56.

Sekuler, R., and Nash, D. 1972. Speed of size scaling in human vision. *Psychonomic Science 27:* 93-94.

Shepard, R. N. 1975. Form, formation, and transformation of internal representations. In R. L. Solso, ed., *Information Processing and Cognition: The Loyola Symposium.* Hillsdale, New Jersey: Lawrence Erlbaum Associates.

Shepard, R. N., and Chipman, S. 1970. Second-order isomorphism of internal representations: Shapes of states. *Cognitive Psychology 1:* 1-17.

Shepard, R. N., and Feng, C. 1972. A chronometric study of mental paper folding. *Cognitive Psychology 3:* 228-243.

Shepard, R. N., and Metzler, J. 1971. Mental rotation of three-dimensional objects. *Science 171:* 701-703.

Shlaer, R. 1971. Shift in binocular disparity causes compensatory change in the cortical structure of kittens. *Science 173:* 638-641.

Spinelli, D. N. 1970a. Recognition of visual patterns. In D. Hamburg, ed., *Perception and its Disorders.* Vol. XLVIII. Research publication of the Association for Research in Nervous and Mental Disease, U.S.A.

Spinelli, D. N. 1970b. OCCAM: A computer model for a content addressable memory in the central nervous system. In K. H. Pribram and D. E. Broadbent, eds., *Biology of Memory.* New York: Academic Press.

Spinelli, D. N., and Hirsch, H. V. B. 1971. Genesis of receptive field shapes in single units of cat's visual cortex. *Federation Proceedings 30:* 615.

Spinelli, D. N., Hirsch, H. V. B., Phelps, R. W., and Metzler, J. 1972. Visual experience as a determinant of the response characteristics of cortical receptive fields in cats. *Experimental Brain Research 15:* 289-304.

Spinelli, D. N., Metzler, J., and Phelps, R. W. 1973. Visual system in early and late infancy. Paper presented at the annual meeting of the Society for Neuroscience. San Diego, November 1973.

Spinelli, D. N., Metzler, J., and Phelps, R. W. 1975. Neural correlates of visual experience in single unit's of cat's visual and somatosensory cortex. Paper presented at the annual meeting of the Society for Neuroscience. New York, November 1975.

Stryker, M. P., and Sherk, H. 1975. Modification of cortical orientation selectivity in the cat by restricted visual experience: A reexamination. *Science 190:* 904-906.

Taylor, R. L. 1972. Reading spatially transformed digits. *Journal of Experimental Psychology 96:* 396-399.

# Cortical Development and Perceptual Invariance

IACQUELINE METZLER*
*Center for Systems Neuroscience*
*University of Massachusetts at Amherst*
D. NICO SPINELLI
*Department of Computer and Information Science*
*and*
*Center for Systems Neuroscience*
*University of Massachusetts at Amherst*

ABSTRACT *Restricting the visual environment of kittens to one or two simple visual patterns has been shown to be a powerful method in studying the development of neuronal connectivity. The results of a recent experiment employing this simple procedure to study the development of tilt constancy are discussed, and a neural net model of the organization and development of the visual cortex is proposed. The model seeks primarily to explain the modification of orientation selectivity and the development of a mechanism to account for orientation or tilt constancy. A possible neurophysiological experiment that would test, in part, the validity of the model is considered.*

## *1. INTRODUCTION*

A continuing effort in the neurosciences is the study of the mechanisms of perception in the hope of further elucidating how sensory experiences are processed and stored in the brain. In recent years, the digital computer, with its extensive

*1975-76 Fellow, presently at the Department of Neurosurgery, Yale University School of Medicine, New Haven, CT.

capabilities, has assisted as well as influenced our search for these neural mechanisms. Theories and neuronal models of brain function can now be tested operationally, either through the development of computer programs that execute a sequence of operations corresponding to the postulated neural functions, or through the simulation of the hypothesized neural networks, thereby bridging further the gap between neurophysiology and psychology.

A problem of particular interest to many investigators studying the visual system is to try to understand the neural basis of the perceptual constancies. Our perception of the visual world remains relatively constant following head or even whole body tilt, something we have all experienced but take for granted. There is some physiological evidence (Horn and Hill, 1969; Spinelli, 1970a; Denney and Adorjani, 1972; Horn, Strechler, and Hill, 1972) that, at least in the cat, this phenomenon of orientation or tilt constancy might be explained by changes in the orientation selectivity of a small number (about five or six percent) of the neurons in the visual cortex during head or body rotation. Apparently, the receptive fields of these cells have dynamic properties.

Owing to our interest in the effects of early experience on the development of the visual system, an experiment which immediately came to mind was to raise kittens with their visual world restricted to a pattern that was gravity stabilized, moving over the retina whenever the animal tilted its head but maintaining a constant orientation in space. This, we felt, would provide a powerful method for studying, at the single-cell level, the neural mechanisms that are basic to the perceptual constancies and, perhaps, even memory phenomena.

## 2. *THE DEVELOPMENT OF TILT CONSTANCY*

In an effort to test the effects of experience on the development of tilt or orientation constancy, eight kittens were placed in a completely dark room before their eyes opened and, beginning at about three and a half weeks of age, received daily selective visual stimulation. Each animal's total visual experience consisted of viewing two vertical bars with one eye and two horizontal bars with the other eye. The stimuli were mounted in goggles each kitten wore whenever it was in the lighted environment. The two bars viewed by the one eye were gravity stabilized so that their orientation remained constant whenever the kitten tilted its head. The orthogonally oriented bars presented to the other eye were fixed so that they always maintained a constant orientation on the retina even when the kitten's head was tilted. In this way, each kitten served as its own control and the two effects could be conveniently compared.

Records were taken from single cells in visual cortex (Area 17) when the kittens were between 12 and 20 weeks old. The receptive field of a cell was mapped with a black spot that moved at a constant velocity by a computer-controlled x-y plotter across a $25^{o}$ by $25^{o}$ white field. Unit activity was recorded at each of 2500 points within this field and displayed on an oscilloscope screen as an *integral contour display* (see Figure 1). With this method, the shape as well as the orientation of the receptive field could be easily visualized.

The shapes of the receptive fields of the neurons recorded from these kittens were classified as either diffuse (55 percent) or elongated (45 percent). The units with elongated receptive fields differed from those found in the normal cat in that (i) they responded to input from only one eye, and (ii) when mapped in the standard position, they had receptive field orientations coincident with the bar pattern to which the eye has been exposed during rearing (cf. Hirsch and Spinelli, 1970, 1971; Spinelli, Hirsch, Phelps, and Metzler, 1972).

Barring any corrective factors, one would expect that when a paralyzed kitten is rotated about its longtiudinal axis the orientation of the receptive field would rotate by the same amount in the same direction, provided, of course, that one makes sure the eye remains stationary in the orbit. The relationship

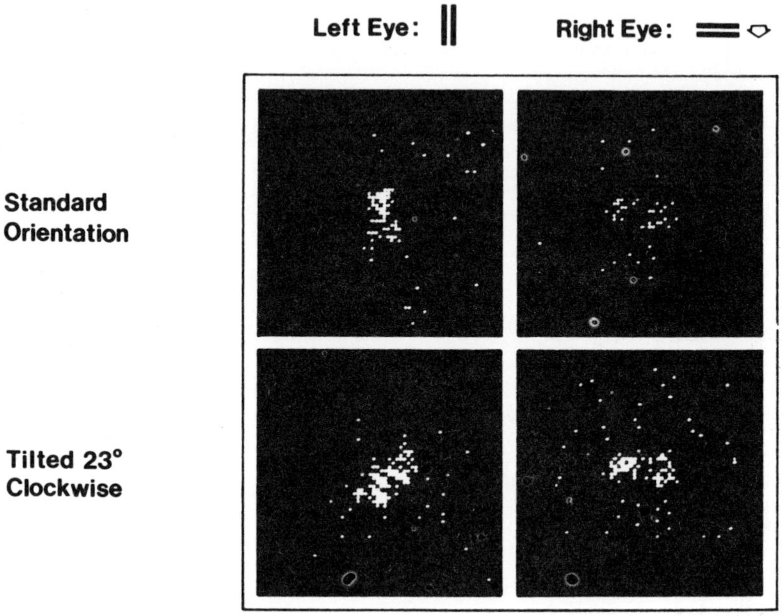


*Figure 1* Integral contour displays of the receptive fields of two cells, each activated by a different eye. These cells were recorded from a kitten whose left eye had been exposed to two vertical bars having a fixed retinal orientation and whose right eye had viewed two gravity-stabilized horizontal bars during development.

between the head axis and the receptive field axis should, therefore, remain constant. This is essentially what we found for the receptive fields of those units that could be activated only by the eye that had viewed the pattern having a fixed retinal orientation during development. Upon returning the animal to the standard position, the field axis also returned to its original orientation. In other words, these receptive fields were not tilt constant. In contrast, the receptive field orientation of cells that could be activated only by the eye that had been exposed to the gravity-stabilized pattern during rearing remained unchanged following body tilt. That is, these cells appeared to demonstrate tilt constancy. The integral contour displays of the receptive fields of two units--one driven by the left eye and the other by the right eye--are shown in Figure 1. These units were recorded from a kitten exposed to two vertical bars having a fixed retinal orientation with the left eye and two gravity-stabilized horizontal bars with the right eye. The receptive fields of both cells were mapped when the kitten was in the standard position and again when the kitten was tilted 23$^{o}$ clockwise. Note that the vertically elongated receptive field of the cell driven by the left eye was not tilt constant, while the orientation of the horizontally elongated receptive field of the cell activated by the right eye remained unchanged following body tilt.*

### 3. *ORIENTATION IN SPACE*

The results of the study described in the preceding section demonstrate that the receptive field orientation of some cells in visual cortex is not immutable. As we have noted, some cortical cells, both in normal animals and in kittens raised under special conditions, appear to compensate for changes in head or body position. Normally, orientation in space depends in large part upon visual cues, but pertinent information is also supplied by input from the vestibular receptors, from proprioceptors in joint capsules which provide data about the relative position of various parts of the body, and from cutaneous exteroceptors, particularly touch and pressure receptors.

Space constancy effects have also been reported in the visual system of invertebrates. Visual receptive fields of certain fibers in the optic nerve of the crayfish are modified by changes in the animal's position in space (Wiersma and Yamaguchi, 1967). Although bilateral destruction of the statocysts appears to abolish the space-constancy properties of these fibers, the authors have not ruled out the possibility that other inputs,

---

*For the details of this experiment, see Metzler and Spinelli, 1977.

including those from joint receptors, may also influence the visual responsiveness of the optic nerve fibers.

The experimental evidence relating to the modification of receptive field orientation during head or body tilt cited above (Horn and Hill, 1969; Spinelli, 1970a; Denney and Adorjani, 1972; Horn et al., 1972; Metzler and Spinelli, 1977) implies a role of visuo-vestibular convergence in perceptual constancy. In addition, it has been clearly demonstrated that activity in the dorsal nucleus of the lateral geniculate body (LGB) and the visual cortex is modified by vestibular stimulation (Grüsser, Grüsser-Cornehls, and Saur, 1959; Grüsser and Grüsser-Cornehls, 1960; Jung, 1961; Jung, Kornhuber, and da Fonseca, 1963; Kornhuber and da Fonseca, 1964; Papaioannou, 1969, 1972, 1973; Magnin, Jeannerod and Putkonen, 1974; Kennedy, Magnin, and Jeannerod, 1976). While the results of these studies indicate that the visual responsiveness of a cell may be altered by prior or concurrent activity in the vestibular pathways, the precise role of this visuo-vestibular integration is not fully understood. Jung et al. (1963) suggest that vestibular afferents to visual cortex function to inform the visual system of body displacements. Perhaps the early stages of such visuo-vestibular convergence may be seen in the properties of some of the cells described in Section 2.

### 4. *PATHWAYS OF VISUO-VESTIBULAR INTEGRATION*

The afferent pathways by which impulses from the labyrinths are relayed to the visual cortex are also poorly defined. Although there is no doubt that there is vestibular modulation of visual response in the LGB (Papaioannou, 1969, 1972, 1973; Magnin et al., 1974; Kennedy et al., 1976), it is unlikely that this vestibular influence is mediated by vestibulo-geniculate pathways since Jung et al. (1963) found activation in the geniculate neurons following labyrinthine polarization only when the posterior cortex and white matter were intact. Thus, it appears that recurrent activation from the cortex is necessary for vestibular modulation of the activity of geniculate neurons. [Cf. Ajmone-Marsan and Morillo (1961), Kalil and Chase (1970), and Sefton and Bruce (1971), demonstrating corticofugal influences upon neurons in LGB. Recent anatomical evidence (Guillery, 1967; Garey, Jones, and Powell, 1968; Niimi, Kawamura, and Ishimaru, 1971) provides structural confirmation of these findings.]

Essentially all neurons in the visual cortex receive specific reticulo-geniculate afferents. While most of these neurons show convergence with nonspecific reticulo-thalamic and vestibular impulses (Grüsser et al., 1959, 1960), a restricted number of geniculate and cortical neurons reveal *specific* convergence of direction specific vestibular and visual input (Grüsser et al., 1959; Jung, 1958, 1961). The long latencies of the reticulo-thalamic and vestibular afferents (Grüsser et al., 1959, 1960; Jung et al., 1963) imply multisynaptic relays, probably located primarily in the brainstem reticular network. The anatomical pathways of the long latency responses and their synaptology in the visual cortex are not well known, although their synaptic termination in the upper cortical layers is consistent with physiological recordings of predominantly surface negative waves after reticulo-thalamic stimulation and axodendritic synapses in the upper layers. There is anatomical (Lorente de Nó, 1928; Szentágothai, 1943) and physiological (Lorente de Nó, 1933; Duensing and Schaefer, 1957; Duensing, 1961) evidence that the reticular formation is the primary relay station of ascending vestibular connections, and vestibular and visual convergence is certain in the paramedian reticular formation. The reticular system is also the premotor coordination center for vestibulo-ocular responses (Lorente de Nó, 1928, 1933; Szentágothai, 1943; Schanzer and Bender, 1959). In the light of these data, it is not unlikely that vestibular influences on the visual system are mediated via the brainstem reticular formation.

## 5. *A MODEL FOR CORTICAL DEVELOPMENT AND PERCEPTUAL INVARIANCE*

We now present a neural net model of the organization and development of the kitten visual cortex, together with a plausible mechanism for tilt constancy. In developing this model, we have attempted to remain in accord with the known anatomical and physiological data. The neuronal circuitry of the model is based on a diagram of intracortical neuronal chains by Lorente de Nó. This diagram, shown in Figure 2, exemplifies the broad plan upon which the central nervous system is organized.

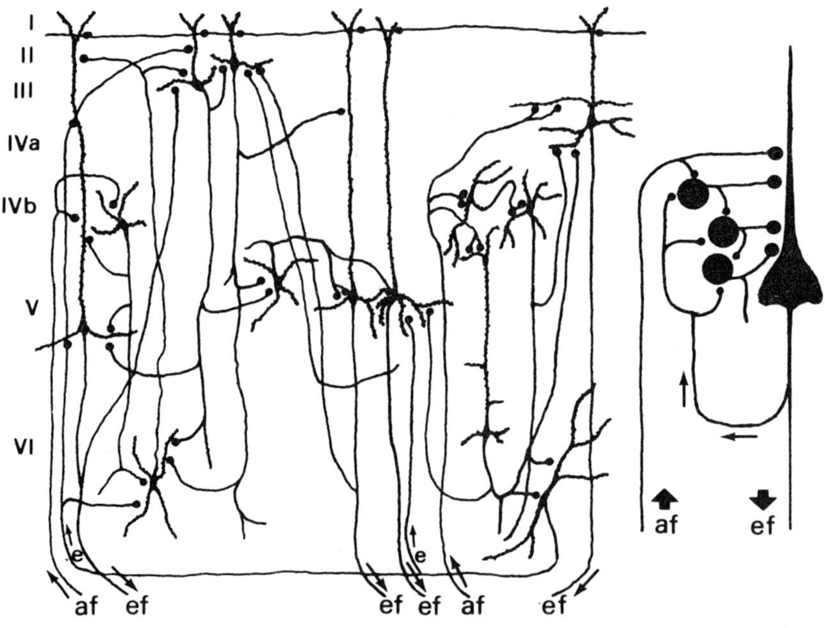


*Figure 2* Diagram of intracortical neuronal chains characteristic of the cerebral cortex. Only a few dendrites and axonal branches are represented. Synaptic junctions are indicated by the darkened circles. Axons entering and leaving the cortex are designated *af* and *ef*, respectively; *e* denotes the recurrent collaterals of cortical internuncial neurons. The diagram at the right is a simplification of the one at the left. Roman numerals at the far left refer to the cortical layers. (After Lorente de Nó, in J. F. Fulton. 1949. *Physiology of the Nervous System,* 3rd ed. New York: Oxford University Press, p. 307.)

### 5.1 *The Elements of the Model*

The intrinsic features of the model are presented in Figure 3, with the symbols used to represent the various elements defined in the legend. The model consists of interconnected networks of cortical cells. The principle cell types are: (i) *Stellate cells* (S), found primarily in layer IV, with short branching axons and numerous dendritic spines; (ii) *Pyramidal cells* (P), in layer V, sending long apical dendrites with many spines toward the cortical surface and axons down into white matter below the cortex; and (iii) *Basket cells of Cajal*, found essentially in all cortical layers with few dendritic spines and an intrinsic distribution of axons.

### 5.2 *Synaptology of the Modeled Cortex*

In "wiring up" the model, we remained as faithful as possible to the following assumptions, most of which are based on anatomical and physiological evidence. The first basic assumption is that cortical cells can be excited and/or inhibited by other cells via excitatory and/or inhibitory synapses. The cells have well-defined thresholds of excitation, and the electrophysiological as well as the chemical time course of excitation is shorter than that of inhibition. Both the suprathreshold excitatory response and the inhibitory response of a cortical neuron can be orientation/direction sensitive (Benevento, Creutzfeldt, and Kuhnt, 1972).

All *extrinsic* input to visual cortex is considered to be excitatory (Garey and Powell, 1971; Benevento et al., 1972). Cortical inhibition is assumed to be *intrinsic* in nature (Benevento, et al., 1972). The dendrites of the stellate cells and pyramids receive excitatory synapses (Colonnier, 1968; Garey, 1971), while the somas of these same neurons receive primarily inhibitory terminals. The specific excitatory input to the cortical neurons is derived from small retinal areas via the dorsal LGB (Benevento et al., 1972), terminating on the dendrites of stellate and pyramidal cells (Colonnier and Rossignol, 1969; Garey and Powell,

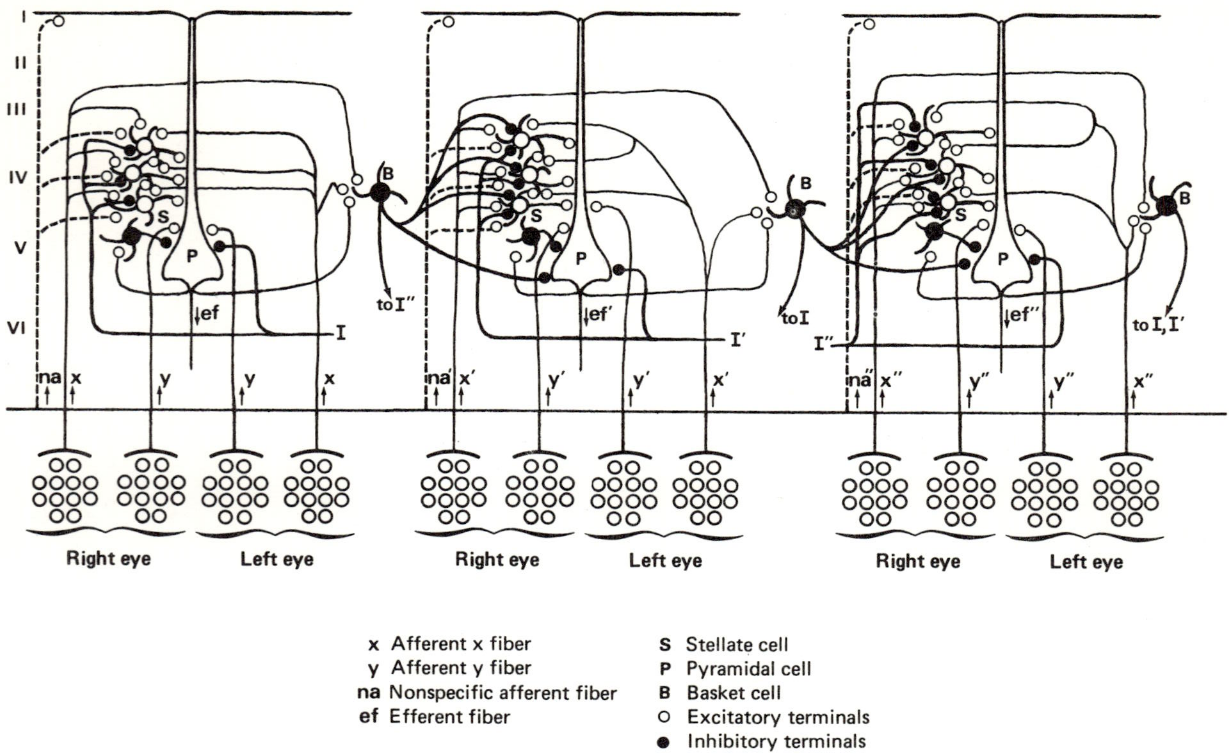


*Figure 3* Diagram of three interconnected networks of cortical cells. The cluster of open circles along the bottom edge of the diagram represents an array of retinal cells having circular receptive fields. Roman numerals at the far left refer to cortical layers. (Cf. Figure 2.)

1971) as well as on the dendrites of intrinsic neurons. The axons of the stellates and the axon collaterals of both the stellate and pyramidal cells terminate as excitatory endings on the dendrites of other cortical neurons (LeVay, 1973). Basket cells distribute axons only within the cortex and terminate as inhibitory synapses on other cortical cells (LeVay, 1973).

Relationships between structure and function have been demonstrated recently by Kelly and Van Essen (1974). They have shown that most "simple" cells are stellate cells, while the majority of "complex" cells are pyramidal neurons. Both "simple" and "complex" cells can receive monosynaptic input from the LGB (Hoffman and Stone, 1971). "Simple" cells are activated more frequently by the small diameter axons of "sustained" or "X" cells in the LGB, while "complex" cells are driven by the larger axons of "transient" or "Y" cells (Stone and Dreher, 1973).

### *5.3 Network Structure and Function*

Let us assume that there exists in the cortex a number of neural networks like those shown in Figure 3. Basically, *each network* consists of a number of stellate cells (S) whose axons terminate in excitatory synapses on a pyramidal cell (P). The pyramidal cell generates an axon which leaves the cortex as a part of the efferent system as well as axon collaterals which make excitatory connections with the basket cells. Afferent "X" and "Y" fibers terminate in excitatory endings on the stellates and pyramids, respectively. The "X" fibers also end on a basket cell. This basket cell, therefore, receives input from both afferent and efferent fibers; in turn, it forms inhibitory connections with the stellate and the pyramidal cells of other networks. It seems natural to assign the function of "comparator" to this interneuron since it "compares" the input and output of each network, signaling the resulting level of correspondence to neighboring networks in the form of inhibition. While there is evidence that many cortical cells receive feed-forward inhibition via disynaptic afferent pathways (Watanabe, Konishi, and Creutzfeldt, 1966), in the model only the "X" fibers terminate on the comparator interneuron. Consequently, the comparator cell signals accurate levels of match for static patterns (i.e., for "X"-cell activity) but less accurate comparisons for transient patterns (i.e., for "Y"-cell activity) because the "Y" fiber input reaches it only indirectly. If, however, a direct anatomical connection from the "Y" cells to the basket cells is discovered, the model will be updated accordingly.

The pyramidal cell is also inhibited by a basket cell within the same network receiving recurrent collaterals from the pyramid. We believe that the function of this inhibitory recurrent pathway is similar to that performed by the Renshaw cell in the spinal cord; i.e., it linearizes and extends the operating range of the pyramidal cell, much like a negative feedback loop in an operational amplifier. The nonspecific afferent fibers (na) terminate on the dendrites of the stellate cells as well as on the branches of the apical dendrites of the pyramidal cell in the upper layers of the cortex.

There are a number of additional assumptions that are crucial to the behavior of the network. First, we assume that the synaptic junctions are modifiable; i.e., they have the characteristic of plasticity. Secondly, synaptic connectivity is considered to be directly proportional to the amount of activity passing through the synaptic junction. In other words, the greater the activity across a synapse, the more conductive the junction will be, and vice versa, such that, *eventually*, the excitatory (or inhibitory) postsynaptic potential becomes a function *not* of the input activity itself, but rather of the previous history of the junction, i.e., the synaptic conductivity. It is further assumed that the comparator interneuron fires only when it receives concurrent input from both the afferent and efferent terminals; the cell will fire maximally when the activity in the afferent and efferent fibers is identical and will fire minimally or not at all when these inputs are completely different (cf. Spinelli, 1970b). In addition, the "X" fibers endow many of the stellate cells with *some* orientation and motion selectivity. Although the pyramidal cells do not possess orientation selectivity as a result of their "Y"-fiber terminals, they may demonstrate a *slight* preference for the orientation(s) of the stellate cells that have excitatory terminals on their apical dendrites. Finally, all networks are considered to be addressed in parallel.

### 5.4 *Network Parameters*

We now consider the three parameters which permit the network to perform the required functions.

i. *Lateral Inhibition.* Initially all the networks receive the same input; i.e., activity arrives at all the networks in parallel and, at first, they all begin to respond to the same stimulus pattern. However, eventually the correspondence between the activity of the input fibers and that of the output of the pyramidal cell in one of the networks will be just a bit better than in any of the neighboring networks. The comparator interneuron associated with this network will be more active and will laterally inhibit the neighboring networks, preventing them from responding to and learning further that stimulus pattern. The number of networks that learn the same pattern is therefore determined by the extent of the lateral spread of the inhibition. In this way, lateral inhibition governs the redundancy of information storage.

ii. *Firing Frequency.* Each time a cortical cell is activated, it responds with a certain rate of discharge. It is assumed that, *initially*, the greater the firing frequency per unit time, the more the synaptic conductivity will be modified (cf. Spinelli, 1970b). Consequently, different amounts of time may be required by different networks to become selectively responsive to various stimulus parameters.

iii. *Input-Output Correspondence*. If the output pattern of activity from the pyramidal cell is identical to that of the input pattern of activity, the level of correspondence registered by the comparator interneuron will be 100 percent, resulting in perfect recognition of the stimulus. The comparator cell, in turn, will fire maximally, inhibiting the neighboring networks. This, of course, is achieved only if the network is subsequently presented with the total stimulus pattern for which it has become highly selective. However, *perfect* repetition of sequences of cellular activity is the exception rather than the rule in the nervous system, owing to such factors as "spontaneous" discharge, "nonspecific" sensory input, as well as a variety of other factors. It is, therefore, highly unlikely that the correspondence between the input and output firing patterns will ever be 100 percent. Further, it is often the case that we can recognize a stimulus pattern with which we are familar when we are presented with only a part of the pattern, and we may even respond appropriately to that stimulus, and perhaps come to expect consequences of our behavior as well (cf. Pribram, Spinelli, and Kamback, 1967). Nevertheless, the smaller the portion of the stimulus presented, the greater is the risk in using it; i.e., the more likely it is that the organism's response will lead to undesirable consequences (see, also, Spinelli, 1970b). Moreover, it is quite likely that a network which is highly selective for a certain pattern has come to *associate* with this stimulus many other bits of information--nonvisual, as well as visual--which, when presented by themselves, may result in partial correspondence between the input and output activities. Ideally, the requisite level of correspondence between input and output patterns of activity should be one that permits the network to respond unmistakably to the stimulus pattern for which it has become highly selective and possibly, on occasion, to those other stimulus features that have become associated with that stimulus (which may be an important factor in memory and imagery), but one that does not introduce a high level of risk into the system.

## 6. *CORTICAL DEVELOPMENT: NEURONAL CONNECTIVITY AND RECEPTIVE FIELD STRUCTURE OF THE MODELED NETWORK*

Let us now consider how each network performs as a function of different amounts as well as various kinds of visual experience.

### *6.1 Visually Naive Animals*

First of all, let us review the structure and function of each network in an organism without visual experience. As we have already noted, afferent fibers of "sustained" or "X" cells in the LGB form excitatory synaptic junctions with the dendrites of the stellate cells. A number of such fibers terminating on a given stellate cell represents a diffuse projection (i.e., the afferent pathways of an array of cells with circular receptive fields) from a particular part of the retina. Since the cortical terminals of these afferent fibers have the characteristic of plasticity, the functional connectivity between the patch of retinal cells and the cortical cell to which they project can be modified. If the synaptic connections of all the afferent projections are strengthened, the receptive field of the postsynaptic cortical cell will be roughly circular or disc-shaped (cf. Spinelli and Barrett, 1969; Pettigrew and Freeman, 1973); whereas, if only the connections from those retinal cells whose circular receptive fields are arranged in a straight line are enhanced, the receptive field of the cortical cell will be line-shaped (cf. Hubel and Wiesel, 1962). Thus, the model allows for a continuum of orientation specificities: the receptive fields of some cortical cells may be very sharply tuned, other fields may be more broadly tuned, while still others may possess no orientation selectivity at all.

Many of the model's stellate cells initially possess *some* orientation and motion selectivity and are only weakly excitatory. The stellates receive input from similarly organized fields or patches in each retina and are, therefore, binocularly activated with varying degrees of monocular dominance.

The afferent fibers from "transient" or "Y" neurons in the LGB terminate in excitatory synapses on the dendrites of the pyramidal cell. These "Y" fibers, like the "X" fibers, are connected to patches of cells in each retina that are similarly organized. Hence, the pyramidal cell is also binocularly activated, but again the inputs from the two eyes may be unequal. The stellate and pyramidal cells in each network receive input from patches of cells in the same general area of each retina; it is the case that these patches may even overlap to a large extent.

While the pyramidal cell possesses some motion sensitivity, the "Y"-fiber terminals do not endow it with orientation selectivity. A pyramidal cell, nevertheless, may demonstrate a *slight*, though rather broad, orientation specificity, corresponding to the orientation(s) of the receptive fields of the stellate cells forming excitatory synaptic junctions with its apical dendrite.

The comparator interneuron provides inhibition between neighboring networks. Initially, the orientation selectivity of the comparator cell is very broad as it receives input from both the "X" fibers and the axon collaterals of the pyramidal cell.

In summary, the stellate cells in the visually naive organism possess some orientation and motion selectivity. The pyramidal and basket cells, on the other hand, demonstrate only slight and very broad orientation tuning. Most cells are binocularly activated with varying degrees of monocular dominance.

### *6.2 Animals With Normal Visual Experience*

Now let us consider what happens under conditions of normal visual experience. A stellate cell receives input from both eyes simultaneously, assuming that the retinal disparity of the two images is appropriate. If the stellate possesses *some* prespecified orientation and motion selectivity, it will respond optimally when a stimulus of the *appropriate* orientation moves across its binocular receptive field in the *preferred* direction. If the stellate cell receives equal input from both eyes, this leads to the development of an effective receptive field in each eye, with a retinal position corresponding to the average disparity of the two images (cf. Shlaer, 1971). However, if the input to the neuron is dominated by one eye, this will result in the formation of an effective receptive field for the dominant eye and a less effective (i.e., smaller and less responsive) receptive field for the nondominant eye.

The direct afferent "Y"-fiber activity to the pyramidal neurons can be reinforced by the excitatory input from the stellate cells, providing that the information carried by the "Y"-fiber input corresponds to the same orientation, retinal position, and disparity as that transmitted by the activity in the stellates. If the above conditions prevail, the pyramidal cell will eventually shift from being relatively nonspecific to having essentially the same preferred orientation as the stellate cells in the same network, despite the fact that the stellate and pyramidal neurons receive essentially separate, though parallel, inputs. If, however, the information conveyed by the "X" and "Y" fibers does not correspond with respect to retinal position and disparity and perhaps to orientation as well, the pyramidal cell may assume varying degrees of orientation specificity, depending on the narrowness or breadth of tuning provided by the stellate cells as well as by its own "Y"-fiber input.

The orientation selectivity of the cells in a particular network can also be improved by intracortical inhibition. Neighboring networks having different orientation sensitivities can inhibit each other laterally, thereby sharpening the response of a given network to an optimally oriented stimulus. If the level of correspondence between the "X"-fiber input and the pyramidal cell output at the comparator interneuron is acceptable, the adjacent networks will be inhibited. If, however, the correspondence between the afferent and efferent activity does not attain an acceptable level, the neighboring networks are not inhibited and all networks will continue to learn the stimulus pattern. The orientation selectivity of the comparator interneuron will reflect the current level of orientation tuning of the network from which it receives its input.

*6.3 Animals Receiving Selective Visual Stimulation*

The model can also explain the results of several experiments in which kittens have received selective visual stimulation.

i. *Monocular Deprivation* (e.g., Wiesel and Hubel, 1963). Cells activated by the stimulated eye do not receive any input from the deprived eye. Therefore, the synapses from the deprived eye will not be strengthened--and may possibly be weakened.

ii. *Binocular Deprivation* (e.g., Wiesel and Hubel, 1965). As neither eye receives stimulation, the neural connections found in the visually naive animal (Section 6.1) will prevail, provided that the period of complete deprivation is not so extensive so as to result in the degeneration of the neural apparatus.

iii. *Artificial Strabismus* (e.g., Hubel and Wiesel, 1965).* The misalignment of the visual axes prevents both eyes from receiving input from corresponding parts of the visual field, unless the angle of disparity produced by the squint is small (cf. Shlaer, 1971). The stimulation received by one eye is not related to that arriving at the other with the exception that, on the average, each eye receives about the same sensory input. Because of the simultaneous but nonsynchronous activation of the cortical neurons, one eye will come to dominate the input to any given network. For cortical cells that are influenced equally by the two eyes, either eye may dominate. Otherwise, it will be the initially nondominant eye that loses its influence entirely.

---

*Under the conditions of artificial strabismus, alternate suture, and incoherent input, a critical factor in the development of synaptic connections appears to be the nonsynergistic use of the two eyes.

iv. *Alternate Suture* (e.g., Hubel and Wiesel, 1965). If each eye is occluded separately (on alternating days, for example), both eyes receive normal stimulation but are never used together. The synaptic connections will develop as they do in the case of artificial strabismus, with one eye coming to dominate a particular network. If, at first, the cortical cells are not influenced equally by the two eyes, it will be the nondominant eye that eventually loses its influence. Otherwise, either eye may dominate.

v. *Modification of Orientation Selectivity by Selective Experience*. Since 1970 several studies of the effect of selective experience on the development or modification of orientation selectivity have been carried out. In some of these investigations, the same visual stimulus was presented to both eyes; in others, each eye experienced a different stimulus pattern.

a. *Corresponding input to both eyes* (e.g., Blakemore and Cooper, 1970). Under these conditions, if the orientation of the stimulus falls within the broad orientation tuning curve of the stellate cells in a given network, the synaptic connections will develop as they do in the normal animal (Section 6.2). If, however, the experienced orientation does not occur within the limits of the orientation sensitivities of the stellates, either the network will not become highly tuned to the stimulus pattern, or if it does, the development of orientation selectivity will proceed more slowly than in the case where the stimulus orientation does correspond to the orientation preferences of the neurons.

b. *Noncorresponding or incoherent input to the two eyes* (e.g., Hirsch and Spinelli, 1970, 1971). In this situation, if either of the experienced orientations happen to be within the boundaries of the orientation preferences of the stellate cells in a particular network, neural connectivity will develop as it does for animals with strabismus (Section 6.3, iii) since the cortical neurons are simultaneously but nonsynchronously activated. The eye experiencing the orientation that most closely corresponds to the broad orientation tuning of the stellates will come to dominate the network. In Hirsch and Spinelli's experiment, it is very likely that either the vertical or the horizontal pattern would fall within the range of orientation sensitivities of many of the neurons. If, on the other hand, *neither* of the stimulus patterns happen to fall within the limits of the orientation preferences of the stellate cells, either the network will not become tuned to either of the experienced orientations, or if it does, development will proceed as it does when the orientation of one of the two incoherent patterns coincides with the orientation sensitivities of the cell but at a slower rate. If the orientation of the stimulus presented to one of the eyes *more closely* corresponds to the broad tuning curve of the stellates, that eye will come to dominate the network.

## 7. *A MECHANISM FOR TILT CONSTANCY*

In Section 1 we noted that following head or body tilt, there are some cells in the visual cortex of the cat that change their receptive field orientation in a direction that would seem to compensate for the change in body position. In addition, we recounted the results of our experiment (Metzler and Spinelli, 1977) in which only the receptive fields of cells driven by the eye viewing a gravity-stabilized pattern were tilt constant, while the receptive fields of those neurons activated by the eye that has been exposed to the pattern having a fixed retinal orientation were not. It was suggested that the tilt-constant properties of some of these neurons might be a function, at least in part, of vestibular afferents that serve to inform the visual cortex about changes in head and body position. We now extend our model of cortical development so as to provide a plausible mechanism for tilt constancy.

One of the basic assumptions of our model is that the cortical synapses are modifiable. Modifiable cortical cells receive input from a patch on the retina consisting of a number of cells with circular receptive fields (see Figure 3). *Orientation specificity* is assumed to be a function of which of the synaptic connections between the retinal patch of cells and the postsynaptic cortical cell are strengthened.

In our model, the *orientation* of an elongated or line-shaped receptive field is governed in part by input from the labyrinths and neck proprioceptors. The interaction between these nonspecific afferents (na) and the visual input occurs at the cortical terminals of the afferent visual fibers (see Figure 3). As the animal is tilted, the information arriving from the vestibular afferents causes the functional connections between the patch of retinal cells and the postsynaptic cortical cell to change; as a result, the orientation of the receptive field relative to the vertical meridian of the retina also changes. The vestibular response, however, is considered capable of modifying the receptive field orientation *only* if the change in the level of activity originating in the labyrinths is coincident with a *corresponding* change in the activity in the visual afferents. To help clarify the function of the vestibular input in controlling receptive field orientation, let us consider the development of receptive field properties in kittens raised in a normal environment.

Although there are neurons in the visual cortex of normally reared cats whose receptive fields are modified by changes in head or body position, these units are encountered rather infrequently: Horn et al. (1969, 1972) reported that the receptive fields of only about 6 percent of the orientation selective cells were modified by tilt; Spinelli (1970b) observed that only about 15 percent of the elongated receptive fields (which themselves made up only 20 percent of all units sampled) were tilt constant; and Denney and Adorjani (1972) found that only about 12 percent of the units they studied had receptive fields that appeared to compensate for changes in head position. Since the orientation of objects in the visual environment remains relatively stable during head or body tilt, why then do these space-constant properties not occur with greater frequency in the normal animal? While we cannot provide a definite answer to this question, we offer the following tentative explanation.

The newborn kitten is rather immobile and its eyes do not open until the seventh to tenth day of life. Even after the time of eye opening, it spends most of its time sleeping and nursing. Images of objects moving over the retina are largely the result of movement in the environment; the kitten at this age still moves very little relative to its surroundings. Consequently, there is no change in the level of activity from the labyrinths, and many units activated by visual stimuli during the first weeks of life become tuned to features of the visual environment without the vestibular component. However, by the time the animal is three and a half to four weeks old--the height of the "critical period" in kittens (Hubel and Wiesel, 1970), it is quite active. During this time, many additional units become selectively tuned to specific visual features. Many of these cells will be activated by visual stimuli in the absence of any head or body movement; some other neurons will be responsive to certain visual features when there is coincident activity arriving via the vestibular afferents. Only these latter units will develop tilt-constant receptive fields. As development proceeds, the response characteristics of the receptive fields of other units are similarly determined such that, in the adult animal, there are many more units that are not modified by head and body tilt than there are units exhibiting tilt-constant properties. Apparently these tilt-constant neurons, though fewer in number, are capable of maintaining the stability of the visual world during changes in head and body position.

Now let us consider the role of the vestibular system in the development of receptive fields of the kitten reared under conditions of controlled visual experience--with one eye exposed to two gravity-stabilized horizontal bars, while the other eye views two vertical bars having a fixed orientation on the retina. This kitten is three and a half weeks old when it first experiences anything visually. At this age it is already very active and changes in head and body position are registered in the activity in the vestibular afferents. The cortical cells activated by the eye viewing the two vertical bars will have vertically oriented receptive fields that *are not* tilt constant (see Figure 1). Since the retinal orientation of the stimulus remains unchanged, the relationship between the field axis and the head axis never changes. Although the vestibular system is activated whenever the kitten tilts its head, there is no corresponding change in activity from the visual afferents which would occur only if the stimulus changes its position on the retina. Therefore, the vestibular connections to this cell will not be strengthened.

The cortical neurons activated by the eye exposed to the two horizontal bars will have horizontally oriented receptive fields that *are* tilt constant (see Figure 1). In this case, the relationship between the head axis and the orientation of the stimulus changes by the same amount but in the opposite direction as the rotation of the head or body. As the cortical cell is registering this change, the vestibular system is also being activated, informing the cell of the degree of head or body tilt. These two effects cancel--much like compensatory eye movements cancel head movement--and the unit will develop a tilt-constant receptive field. Just how sharply tuned or how accurate the orientation of the receptive field is in the tilted position depends on the *specificity* of the information arriving from the labyrinths.

Moreover, given the assumption that *most* cortical cells are activated by stimulation of the labyrinths, the model predicts that since there is *no* corresponding change in activity in the visual afferents from the eye exposed to the pattern having a fixed orientation on the retina during head tilt, *no* cortical neurons driven by this eye should be tilt constant. On the other hand, since *all* of the cortical cells activated by the eye that had viewed the gravity-stabilized pattern register a change in activity originating in the retinal fibers during head tilt, and since *most* (but not all) of these cells are also activated by vestibular afferents, it follows that *most* (but not all) of these cells should be tilt constant. The results of our experiment (Metzler and Spinelli, 1977) are consistent with the predictions of the model.

## *8. CONCLUSIONS*

The simulation of the network described above is currently being carried out in collaboration with Bryant York in an attempt to verify whether or not the model can perform the functions specified. The results of this simulation will be the subject of a later report.

Another approach to establishing the validity of the model would be to carry out neurophysiological investigations to test the hypothesized functions. Certainly one of the first things that should be established is whether the synaptic terminals of fibers conveying information from the labyrinths are indeed plastic. This assumption could be easily tested by performing an experiment analogous to the monocular suture study of Wiesel and Hubel (1963) which demonstrated that short periods of monocular experience produced a dramatic reduction of cells activated by the sutured eye. As it has been reported that substantial effects can be obtained following very brief periods of unilaterally restricted experience (Hubel and Wiesel, 1970; Blakemore and Mitchell, 1973), one could use a long-acting local anesthetic--

e.g., Zyljectin which, when used with human patients, remains active for several days--to unilaterally block the activity of the eighth nerve for a few days at the height of the "critical period." The expectation would be that cells normally connected to both labyrinths would retain connections only to the functioning labyrinth. More sophisticated experiments could then be designed to test more detailed aspects of the network.

The model described in the preceding sections is characterized by many highly specific features. To the extent that the hypotheses and assumptions of the model can be verified physiologically and by computer simulation, it may contribute greatly to our understanding of neural mechanisms of pattern recognition, the perceptual constancies, and perhaps even higher level cognitive functions.

ACKNOWLEDGMENTS *This research was supported in part by a grant from the Alfred P. Sloan Foundation to the Center for Systems Neuroscience, and in part by an NIMH Postdoctoral Fellowship 1 F01 MH44282-01 to the first author and by an NIMH Grant No. 7 R01 MH25329 to the second author.*

*REFERENCES*

Ajmone-Marsan, C., and Morillo, A. 1961. Cortical control and callosal mechanisms in the visual system of the cat. *Electroencephalography and Clinical Neurophysiology 13:* 553-563.

Benevento, L. A., Creutzfeldt, O. D., and Kuhnt, U. 1972. Significance of intracortical inhibition in the visual cortex. *Nature, New Biology 238:* 124-126.

Blakemore, C., and Cooper, G. F. 1970. Development of the brain depends on the visual environment. *Nature 228:* 477-478.

Blakemore, C., and Mitchell, D. E. 1973. Environmental modification of the visual cortex and the neural basis of learning and memory. *Nature 241:* 467-468.

Colonnier, M. L. 1968. Synaptic patterns on different cell types in the different laminae of the cat visual cortex: An electron microscopy study. *Brain Research 9:* 268-287.

Colonnier, M. L., and Rossignol, S. 1969. Heterogeneity of the cerebral cortex. In H. H. Jasper, A. A. Ward, and A. Pope, eds., *Basic Mechanisms of the Epilepsies*. Boston: Little, Brown and Company.

Denney, D., and Adorjani, C. 1972. Orientation specificities of visual cortical neurons after head tilt. *Experimental Brain Research 14:* 312-317.

Duensing, F. 1961. Die Erregungskonstellationen im Rautenhirn des Kaninchens bei den Labyrinthstellreflexen (Magnus). *Naturwissenschaften 48:* 681-690.

Duensing, F., und Schaefer, K.-P. 1957. Die Neuronenaktivität in der Formatio reticularis des Rhombencephalons beim vestibulären Nystagmus. *Archiv für Psychiatrie und Nervenkrankheiten 196:* 265-290.

Fulton, J. F. 1949. *Physiology of the Nervous System,* 3rd ed. New York: Oxford University Press.

Garey, L. J. 1971. A light and electron microscopic study of the visual cortex of the cat and monkey. *Proceedings of the Royal Society B 179:* 21-40.

Garey, L. J., Jones, E. G., and Powell, T. P. S. 1968. Interrelationships of striate and extrastriate cortex with the primary relay sites of the visual pathway. *Journal of Neurology, Neurosurgery and Psychiatry 31:* 135-157.

Garey, L. J., and Powell, T. P. S. 1971. An experimental study of the termination of the lateral geniculo-cortical pathway in the cat and monkey. *Proceedings of the Royal Society B 179:* 41-63.

Grüsser, O. J., und Grüsser-Cornehls, U. 1960. Mikroelektrodenuntersuchungen zur Konvergenz vestibulärer und retinaler Afferenzen an einselnen Neuronen des optischen Cortex der Katze. *Pflügers Archiv für die Gesamte Physiologie des Menchen und der Tiere 270:* 227-238.

Grüsser, O. J., Grüsser-Cornehls, U., und Saur, G. 1959. Reaktionen einzelner Neurone im optischen Cortex der Katze nach elektrischer Polarisation des Labyrinths. *Pflügers Archiv für die Gesamte Physiologie des Menchen und der Tiere 269:* 593-612.

Guillery, R. W. 1967. Patterns of fiber degeneration in the dorsal lateral geniculate nucleus of the cat following lesions in the visual cortex. *Journal of Comparative Neurology 130:* 197-222.

Hirsch, H. V. B., and Spinelli, D. N. 1970. Visual experience modifies distribution of horizontally and vertically oriented receptive fields in cats. *Science 168:* 869-871.

Hirsch, H. V. B., and Spinelli, D. N. 1971. Modification of the distribution of receptive field orientation in cats by selective visual exposure during development. *Experimental Brain Research 13:* 509-527.

Hoffman, K.-P., and Stone, J. 1971. Conduction velocity of afferents to cat visual cortex: A correlation with cortical receptive field properties. *Brain Research 32:* 460-466.

Horn, G., and Hill, R. M. 1969. Modifications of receptive fields of cells in the visual cortex occurring spontaneously and associated with bodily tilt. *Nature 221:* 186-188.

Horn, G., Strechler, G., and Hill, R. M. 1972. Receptive fields of units in the visual cortex of the cat in the presence and absence of bodily tilt. *Experimental Brain Research 15:* 113-132.

Hubel, D. H., and Wiesel, T. N. 1962. Receptive fields, binocular interaction and functional architecture in the cat's visual cortex. *Journal of Physiology 160:* 106-154.

Hubel, D. H., and Wiesel, T. N. 1965. Binocular interaction in striate cortex of kittens reared with artificial squint. *Journal of Neurophysiology 28:* 1041-1059.

Hubel, D. H., and Wiesel, T. N. 1970. The period of susceptibility to the physiological effects of unilateral eye closure in kittens. *Journal of Physiology 206:* 419-436.

Jung, R. 1958. Coordination of specific and nonspecific afferent impulses at single neurons of the visual cortex. In H. H. Jasper et al., eds., *Reticular Formation of the Brain.* Boston: Little, Brown and Company.

Jung, R. 1961. Neuronal integration in the visual cortex and its significance for visual information. In W. A. Rosenblith, ed., *Sensory Communication.* London: MIT Press.

Jung, R., Kornhuber, H. H., and da Fonseca, J. S. 1963. Multisensory convergence on cortical neurons. In G. Moruzzi, A. Fessard, and H. H. Jasper, eds., *Progress in Brain Research,* Vol. 1. Amsterdam: Elsevier.

Kalil, R. E., and Chase, R. 1970. Corticofugal influence on activity of lateral geniculate neurons in the cat. *Journal of Neurophysiology 33:* 459-474.

Kelly, J. P., and Van Essen, D. C. 1974. Cell structure and function in the visual cortex of the cat. *Journal of Physiology 238:* 515-547.

Kennedy, H., Magnin, M., and Jeannerod, M. 1976. Receptive field response of LGB neurons during vestibular stimulation in awake cats. *Vision Research 16:* 119-120.

Kornhuber, H. H., and da Fonseca, J. S. 1964. Optovestibular integration in the cat cortex: A study of sensory convergence on cortical neurons. In M. B. Bender, ed., *The Oculomotor System*. New York: Hoeber Medical Division, Harper and Row.

LeVay, S. 1973. Synaptic patterns in the visual cortex of the cat and monkey. Electron microscopy of Golgi preparations. *Journal of Comparative Neurology 150:* 53-86.

Lorente de Nó, R. 1928. *Die Labyrinthreflexe auf die Augenmuskeln nach einseitiger Labyrinthexstirpation nebst einer kurzen Angabe über den Nervenmechanismus der vestibulären Augenbewegungen*. Wien: Urban und Schwarzenberg.

Lorente de Nó, R. 1933. Vestibulo-ocular reflex arc. *Archives of Neurology and Psychiatry 30:* 245-291.

Magnin, M., Jeannerod, M., and Putkonen, P. T. S. 1974. Vestibular and saccadic influences on dorsal and ventral nuclei of the lateral geniculate body. *Experimental Brain Research 21:* 1-18.

Metzler, J., and Spinelli, D. N. 1977. Tilt constancy mechanisms in the developing visual cortex (in preparation).

Niimi, K., Kawamura, S., and Ishimaru, S. 1971. Projections of the visual cortex to the lateral geniculate and posterior thalamic nuclei in the cat. *Journal of Comparative Neurology 143:* 279-312.

Papaioannou, J. 1969. Vestibular influences on the spontaneous activity of neurones in the lateral geniculate nucleus of the cat. *Journal of Physiology 202:* 86P.

Papaioannou, J. N. 1972. Electrical stimulation of vestibular nuclei: Effects on light evoked activity of lateral geniculate neurones. *Pflügers Archiv für die Gesamte Physiologie des Menchen und der Tiere 334:* 129-140.

Papaioannou, J. N. 1973. Changes in the light evoked discharges from lateral geniculate nucleus neurones in the cat, induced by caloric labyrinthine stimulation. *Experimental Brain Research 17:* 10-17.

Pettigrew, J. D., and Freeman, R. D. 1973. Visual experience without lines: Effect on developing cortical neurons. *Science 182:* 599-601.

Pribram, K. H., Spinelli, D. N., and Kamback, M. C. 1967. Electrocortical correlates of stimulus response and reinforcement. *Science 157:* 94-96.

Sefton, A. J., and Bruce, I. S. C. 1971. Properties of cells in the lateral geniculate nucleus. *Vision Research Supplement* (3): 239-252.

Shanzer, S., and Bender, M. B. 1959. Oculomotor responses on vestibular stimulation of monkeys with lesions of the brainstem. *Brain 82:* 669-682.

Shlaer, R. 1971. Shift in binocular disparity causes compensatory change in the cortical structure of kittens. *Science 173:* 638-641.

Spinelli, D. N. 1970a. Recognition of visual patterns. In D. Hamburg, ed., *Perception and its Disorders,* Vol. XLVIII. Research publication of the Association for Research in Nervous and Mental Disease, U.S.A.

Spinelli, D. N. 1970b. OCCAM: A computer model for a content addressable memory in the central nervous system. In K. H. Pribram and D. E. Broadbent, eds., *Biology of Memory*. New York: Academic Press.

Spinelli, D. N., and Barrett, T. W. 1969. Visual receptive field organization of single units in the cat's visual cortex. *Experimental Neurology 24:* 76-98.

Spinelli, D. N., Hirsch, H. V. B., Phelps, R. W., and Metzler, J. 1972. Visual experience as a determinant of the response characteristics of cortical receptive fields in cats. *Experimental Brain Research 15:* 289-304.

Stone, J., and Dreher, B. 1973. Projection of X- and Y-cells of the cat's lateral geniculate nucleus to areas 17 and 18 of visual cortex. *Journal of Neurophysiology 36:* 551-567.

Szentágothai, J. 1943. Die zentrale Innervation der Augenbewegungen. *Archiv für Psychiatrie und Nervenkrankheiten 116:* 721-760.

Watanabe, S., Konishi, M., and Creutzfeldt, O. D. 1966. Postsynaptic potentials in the cat's visual cortex following electrical stimulation of afferent pathways. *Experimental Brain Research 1:* 272-283.

Wiersma, C. A. G., and Yamaguchi, T. 1967. Integration of visual stimuli by the crayfish central nervous sytem. *Journal of Experimental Biology 47:* 409-431.

Wiesel, T. N., and Hubel, D. H. 1963. Single-cell responses in striate cortex of kittens deprived of vision in one eye. *Journal of Neurophysiology 26:* 1003-1017.

Wiesel, T. N., and Hubel, D. H. 1965. Comparison of the effects of unilateral and bilateral eye closure on cortical unit responses in kittens. *Journal of Neurophysiology 28:* 1029-1040.

# A Neural Model for the Handling of Phenomena Associated with Trains of Light Stimuli: An Updated Version to Fit Fusion Data

SHUN-ICHI AMARI*
*Center for Systems Neuroscience*
*University of Massachusetts at Amherst*

ISRAEL LIEBLICH†
*Center for Systems Neuroscience*
*University of Massachusets at Amherst*

ARTHUR I. KARSHMER
*Department of Computer and Information Science*
*and*
*Center for Systems Neuroscience*
*University of Massachusetts at Amherst*

ABSTRACT *A computer model is presented which explains most of the strong phenomena related to the presentation of short and long trains of light stimuli. The cases of masking, metacontrast, and flicker fusion are discussed.*

## 1. *INTRODUCTION*

In successive chapters of the *Handbook of Sensory Physiology* (Jameson and Hurvich, 1972), Weisstein and Kelly discussed the phenomena of light-train stimulation from two different points of view--Weisstein on metacontrast and Kelly on flicker. While both

---

*1975-76 Fellow, on leave from the Department of Mathematical Engineering and Instrumentation Physics, University of Tokyo, Tokyo, Japan.

†1975-76 Fellow, on leave from the Department of Psychology, The Hebrew University of Jerusalem, Jerusalem, Israel.

authors treat their major findings and models relating to their specific area of interest in depth, it is not obvious to the reader that the two phenomena are directly related. In particular, Weisstein's use of the Landahl-Rashevsky model and Kelly's use of an N-stage filter model do little to show the relationships between metacontrast and flicker fusion. In a chapter by Boynton, however, entitled "Discrimination of Homogeneous Double Pulses of Light," the author suggests that there is a possible relation between the two areas.

The present paper proposes an updated, simple neural model (cf. Amari, Lieblich, and Karshmer, 1976) which summarizes most of the strong phenomena related to trains of light stimuli. This model also fits phenomena related to the presentation of a single stimulus--e.g., Mach bands (Ratliff, 1965).

For convenience of presentation, we dichotomize the phenomena observed into a category related to short trains of light stimuli and a category related to long trains of light stimuli.

### *1.1 Short Trains of Light Stimuli*

i. One of the intriguing results in the investigation of perception of a small number of pulses in a flicker train is that the shorter the flash train, the larger must be the interval between pulses to perceive flicker (Boynton, 1972).

ii. Masking and metacontrast effects (reviewed in Kahneman, 1968; Weisstein and Rashevsky, 1968; Boynton, 1972; Weisstein, 1972; Weisstein, Ozog, and Szoc, 1975) occur when two stimuli (or more, in some cases) are presented close to each other in time and/or in space. In the masking paradigm, one stimulus (target-stimulus, TS) may be a brief flash illuminating a given retinal target, while the second stimulus (masking-stimulus, MS) may have characteristics the same as or different than the TS in terms of duration, intensity, and area. The MS may coincide with the TS in onset time or may be flashed with some onset asynchrony. The MS illuminates at least the retinal area illuminated by the TS.

### *1.2 Long Trains of Light Stimuli*

The strongest phenomena related to long trains of flashes are the following (Kelly, 1972):

i. The Ferry-Porter Law: Over a considerable range, critical flicker fusion (CFF, the threshold for reporting fusion of flickering stimuli by observers) varies approximately as the logarithm of the stimulus amplitude.

ii. The Talbot-Plateau Law: The apparent brightness of the periodic stimulus closely matches that of a steady stimulus of the same average luminance.

iii. The Granit-Harper Law: Fusion frequency increases with the area of the flickering field (Woodworth, 1939).

iv. The modulation index versus intensity is a curvilinear relation with a minimum at some medium intensity (Kelly, 1972).

In the metacontrast paradigm, the TS and the MS are adjacent to one another in space but not superimposed. The strong result related to masking (Kahneman, 1968) is the relation of the effect to stimulus onset asynchrony (SOA) between the TS and the MS. When the TS precedes the MS by 50 to 100 msec, the threshold for TS detection rises steeply. Masking effects are usually most severe when SOA is about 0.0. The threshold for detection of the TS gradually declines with further decreases in SOA until it finally reaches a light adapted value. For the metacontrast case, when the target-to-mask energy ratio is close to 1.0, an inverted U-shaped function of the masking effect with SOA is obtained, the maximum effect being observed at around 50 msec SOA. In metacontrast, the detection of the TS is not impaired. Only its brightness characteristics change (Weisstein, 1972).

At this stage, an excellent five-neuron model was presented by Weisstein (Weisstein and Rashevsky, 1968; Weisstein, 1972; Weisstein et al., 1975) which accounts for the most important metacontrast data. Several models have been suggested for flicker phenomena (see Kelly, 1972, for an evaluation of the models). Weisstein's model does not deal explicitly with masking phenomena, and although nonlinearities are suggested, it would not predict the results observed in experiments employing long trains of stimuli.

The proposed model is an extension of the ideas on which Weisstein's model rests to a wider range of phenomena. In this sense, it is also a demonstration of the way in which model building should proceed. Models should grow hierarchically, describing a growing perimeter of phenomena, rather than becoming too narrow by trying immediately to take into account all the reliable and unreliable data of a very specialized area.

## 2. *THE MODEL*

The model is a network composed of three layers of neurons (Figure 1). One is a layer of excitatory interneurons and another is a layer of inhibitory interneurons. It is assumed that light

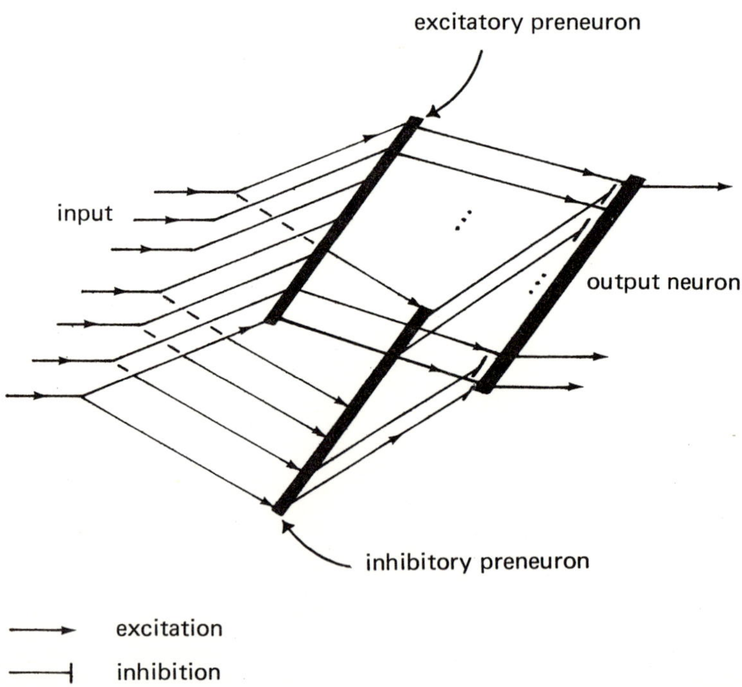


*Figure 1*

stimuli excite the corresponding positions of both excitatory and inhibitory interneurons at the same time. The output of these neurons, in turn, is combined to facilitate or inhibit the neurons of the third layer, which we call output neurons, the inhibitory connections being of the forward-lateral type. Since forward inhibition is easily transformable into recurrent inhibition in the linear case, we do not feel that it will substantially limit the generality in the model's results.

A single neuron's behavior is described by the equations:

$$\tau\dot{u}(t) = - u + s(t),$$
$$x(t) = f[u(t)],$$

where $u(t)$ is the membrane potential, $s(t)$ is the weighted sum of input stimuli to the neuron, and $x(t)$ is the output pulse rate at time $t$. $\tau$ is a time constant which represents the rapidity of response of the neuron, and $f$ is a nonlinear function relating the membrane potential $u$ to the output $x$.

The anatomy of the network and its parameters are determined by the following considerations. First, in order to fit Mach band phenomena, the model should have lateral-inhibitory properties. For this purpose, we chose a wide fan-out connection from inhibitory interneurons to the output neurons, whereas excitatory interneurons have a relatively narrow fan-out.

Secondly, in order to fit the masking effect, the time constant $\tau_{in}$ of an inhibitory neuron should be much smaller than the time constant $\tau_{ex}$ of an excitatory neuron. In this case, inhibition arrives first and excitation arrives later at an output neuron for a given pulse of light. For a train of two light pulses, the early onset of the inhibition from the second stimulus reduces the delayed excitation from the first stimulus, thus effecting backward masking. Metacontrast could be explained by the combination of the characteristics of spatial lateral inhibition and the difference of temporal characteristics of excitatory and inhibitory interneurons.

Flicker fusion itself is rather easy to explain because the temporal behavior of our neurons plays the role of a low-pass filter. A high-frequency pulse train will be fused into a rippling wave after passing through a low-pass filter. It is difficult, however, to explain the Ferry-Porter effect using such a linear model because when the strength of stimuli is doubled, the outputs of a linear model are also doubled. Thus, one cannot expect changes in modulation of the output with changes in the intensity of stimuli for a given frequency. The Ferry-Porter Law can best be explained by introducing nonlinearity (saturation) into the output function f(u). It is assumed that the output of an excitatory neuron saturates more rapidly and that the output of an inhibitory neuron saturates at a slower rate, maintaining a comparatively prolonged linear component. The strength of excitation becomes only slightly larger, therefore, when there is a doubling of the strength of stimuli concurrent with the output of the excitatory neurons being saturated. However, the strength of inhibition becomes nearly doubled if inhibitory neurons are not yet saturated. In this case, the increase in the strength of stimuli causes an increase in the inhibition, and changes a continuously rippling output wave of the output neurons into an intermittent flickering one, thus explaining the Ferry-Porter effect. In order to produce the modulation index versus intensity relation, an excitatory neuron should have a more rapidly increasing nonlinear function for small u than an inhibitory neuron.

### 3. *COMPUTER REALIZATION OF THE MODEL*

The following simplifications were made for the computer model:

i. The layers were conceived as an array of modules composed of one of each of the three types of neurons. To simulate the backward masking and flicker-fusion phenomena, it was necessary to "record" from the modules which were located at the point of light stimulation. Further, by treating the stimulated modules as a single entity, we were able to construct a single module model which was sufficient to realize most of the backward masking and flicker phenomena, with the constraint that when the size of the stimulation area changes, there must also be a corresponding change in the connection parameters. In order to realize the metacontrast phenomenon, an additional three-neuron module was attached to the previous module. The new module was attached by way of forward-lateral inhibition (the two inhibitory interneurons being connected to the two output neurons, each excitatory interneuron being connected only to its output neuron).

ii. The nonlinearity function assumed the following form:

$$f(u) = a\,F(cu - d) - b, \qquad (u \geq 0)$$

where

$$F(u) = \frac{1}{\sqrt{2\pi}} \int_{-\infty}^{u} \exp\left\{-\frac{z^2}{2}\right\} dz$$

is the integral of the standardized normal distribution. The model was developed on a GT-44 graphics computer system, which is based on the DEC PDP-11/40 processor.

For its inputs, the program asks the user the following question: Kind of experiment (mask, metacontrast, or fusion)? (In case of fusion, a "narrow" and "wide" case is optional for observing the Granit-Harper effect.) Given the response, the program asks for relevant durations and stimulus intensities, as well as the degree of light adaptation desired. It then asks for the synaptic weights and saturation values for the different neurons in the model.

Using an experimental version of the model, we were able to fix the following parameters:

i. Time constants:

$$\tau_{ex} = 100 \text{ msec},$$

$$\tau_{in} = 25 \text{ msec},$$

$$\tau_{out} = 25 \text{ msec}.$$

ii. All synaptic weights are set equal to 1 except for the lateral-inhibitory connection which is given the value of -0.3.

iii. Nonlinearity constants:

$$\left.\begin{aligned} a &= 2.4 \\ b &= 0.8 \\ c &= 2.5 \\ d &= 1.0 \end{aligned}\right\} \text{ for excitatory neurons,}$$

$$\left.\begin{aligned} a &= 7.0 \\ b &= 0.56 \\ c &= 0.5 \\ d &= 0.2 \end{aligned}\right\} \text{ for inhibitory neurons.}$$

## 4. *RESULTS OF EXPERIMENTS*

Figure 2 shows the variation of the masking effect with SOA (complete dark adaptation--ambient light = 0.0, TS/MS energy ratio = 1.00).

The masking effect is defined as

$$\frac{\text{total excit. by TS} - \text{total excit. by TS followed by MS}}{\text{total excit. by TS}},$$

with total excitation being defined as the summation of the excitation of output neurons over time until excitation decays to 0.5 of the maximal activity of the neuron.

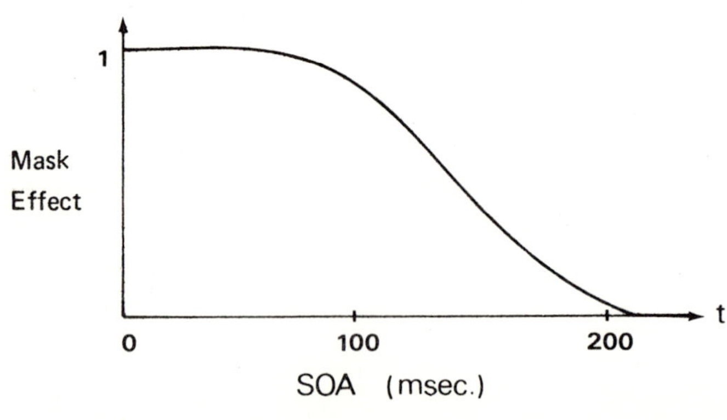


*Figure 2*

Figure 3 shows the variation of masking with SOA in the metacontrast case (ambient light = 0.0, TS/MS energy ratio = 1.00). The masking effect is defined as above.

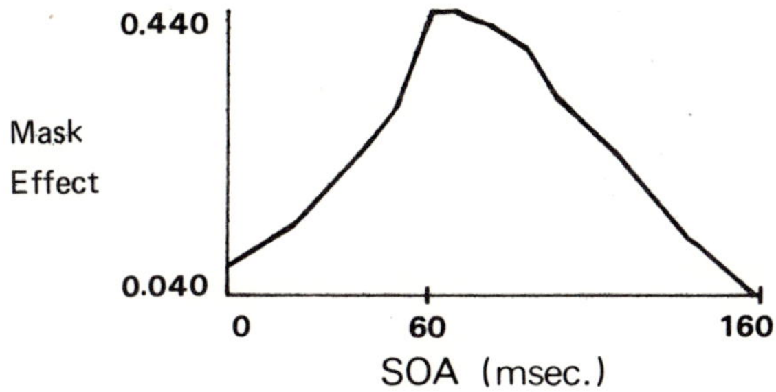


*Figure 3*

Figure 4 shows the results of fusion experiments varying intensities from 0.1 to 0.7 and frequencies from 7 to 40 Hz.

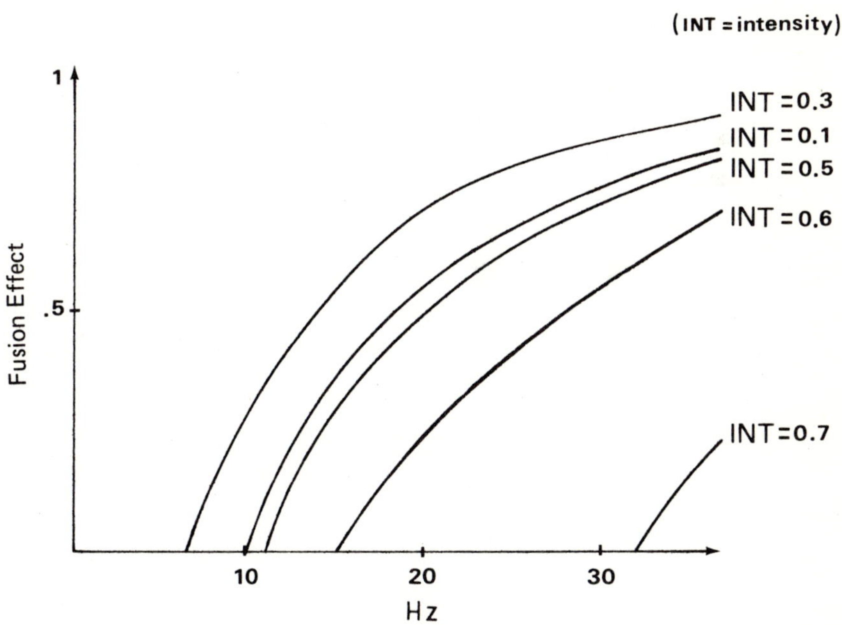


*Figure 4*

The fusion effect is defined as

$$\frac{\text{maximum excitation - minimum excitation}}{\text{maximum excitation}}$$

after a sufficient number of light stimuli have been applied.

Figure 5 shows the relation between the fusion effect and the intensity of light for a fixed frequency (25 Hz) light stimulus. The modulation index versus intensity relation can be clearly observed.

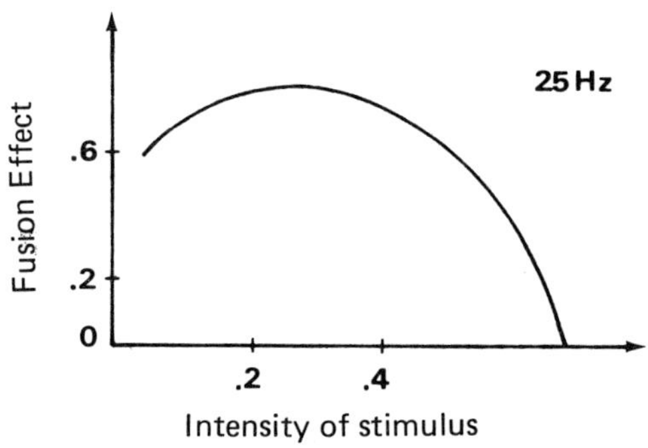


*Figure 5*

## 5. *DISCUSSION*

It is clear that, in general, most of the aforementioned strong phenomena can be realized with this model, e.g., the backward masking effects, the inverted U function of the metacontrast effect, and the Ferry-Porter relationship. In addition, we could approximate the Talbot-Plateau results and the Granit-Harper effect.

We also observed that the model tended to produce the effects of short trains of flashes in the paradigm (Boynton, 1972). These effects are a direct "post-diction" of the model, since we did not attempt initially to use these effects in defining the model. As mentioned above, our definition of fusion takes into account effects of short light flashes in that it is computed after a sufficient number of flashes have been applied to the model.

We feel that further research using this model could be carried out in the following areas:

i. Experimenting further with the one-dimensional or two-dimensional model.

ii. Refining the form of nonlinearity of the model. This is especially important in the case of the fusion phenomena. A sigmoid-type function seems to be appropriate to account for the modulation sensitivity data (Kelly, 1972). It is of interest to note that the masking and metacontrast phenomena would not be greatly affected by this change. The addition of the flicker phenomena to the masking model enriches the predictions of the characteristics of the neurons in a net which has to perform *all* the effects.

It is interesting to compare the current model with the Yasuda-Hiwatashi three-layer model (Yasuda and Hiwatashi, 1970; Yasuda, 1971) in which the authors showed the on-response and off-response for an on-center x-type cell. The current model is able to produce similar results with the proper parameters, though the character of the time lags is different. Note, also, that the current model shows an off-response to a very strong stimulus.

It seems to us that the power of the model rests on:

i. Its extreme simplicity.

ii. The wide range of phenomena covered by it--i.e., Mach bands, masking, and flicker phenomena.

## 6. *PREDICTIONS*

i. Some form of enhancement of the MS is expected to appear in masking experiments, and this tendency should decline with SOA. Enhancement of the MS is not expected in the metacontrast paradigm.

ii. Processes in the retinal system showing both a time constant on the order of 100 msec and rapid saturation could possibly be identified as an excitatory process.

iii. Processes in the retinal system showing both a time

constant on the order of 20 msec and slow saturation could possibly be identified as an inhibitory process.

## *REFERENCES*

Amari, S., Lieblich, I., and Karshmer, A. 1975. A neural model for the handling of phenomena associated with trains of light stimuli. *Brain Theory Newsletter 1*(2): 20-23.

Boynton, R. M. 1972. Discrimination of homogeneous double pulses of light. In D. Jameson and L. M. Hurvich, eds., *Handbook of Sensory Physiology*. Vol. 7. *Visual Psychophysics*. Berlin: Springer-Verlag.

Jameson, D., and Hurvich, L. M., eds. 1972. *Handbook of Sensory Physiology*. Vol. 7. *Visual Psychophysics*. Berlin: Springer-Verlag.

Kahneman, D. 1968. Method, findings, and theory in studies of visual masking. *Psychological Bulletin 70:* 404-425.

Kelly, D. H. 1972. Flicker. In D. Jameson and L. M. Hurvich, eds., *Handbook of Sensory Physiology*. Vol. 7. *Visual Psychophysics*. Berlin: Springer-Verlag.

Ratliff, F. 1965. *Mach Bands: Quantitative Studies on Neural Networks in the Retina*. San Francisco: Holden-Day.

Weisstein, N. 1972. Metacontrast. In D. Jameson and L. M. Hurvich, eds., *Handbook of Sensory Physiology*. Vol. 7. *Visual Psychophysics*. Berlin: Springer-Verlag.

Weisstein, N., Ozog, G., and Szoc, R. 1975. A comparison and elaboration of two models of metacontrast. *Psychological Review 82:* 325-343.

Weisstein, N., and Rashevsky, A. 1968. Landahl neural net: simulation of metacontrast. *Psychological Review 75:* 494-521.

Woodworth, R. S. 1939. *Experimental Psychology*. New York; Holt, Rinehart and Winston.

Yasuda, M. 1971. A dynamic model of the vertebrate retina. *Kybernetik 9:* 26-30.

Yasuda, M., and Hiwatashi, K. 1970. Spatio-temporal information processing in retina model. *Systems-Computers-Controls 1:* 9-16.

# A Mathematical Approach to Neural Systems

SHUN-ICHI AMARI*
*Center for Systems Neuroscience*
*University of Massachusetts at Amherst*

ABSTRACT *The brain is indeed a highly hierarchical system in which a component net of one level functions as the building block of higher-level nets. It is necessary to develop a proper mathematical method and technique for elucidating each level. The present rather expository paper summarizes the author's recent endeavor of developing various mathematical techniques and obtaining characteristic features of nerve nets of the various levels. Following the Introduction (Section 1), Section 2 treats the dynamics of neuron pools, which are the elementary building blocks of neural systems. A simple neuron pool has monostable or bistable dynamics depending on the parameters of the pool. The dynamical properties of a neuron pool give rise to multistable, hysteretic, and oscillatory phenomena, which play an essential role in a higher-level system. In Section 3 we consider the dynamics of neural fields, which can be regarded as systems consisting of spatially distributed neuron pools having mutual connections. We present a mathematical method by which we analyze the dynamics of pattern formation in neural fields. Section 4 deals with the learning or self-organizing capabilities of neural systems. We search for a common neural mechanism which realizes high-level information processing, such as association and concept formation.*

---

*1975-76 Fellow, on leave from the Department of Mathematical Engineering and Instrumentation Physics, University of Tokyo, Tokyo, Japan.

## 1. *INTRODUCTION*

There are many methods of approach to the theoretical study of brain functioning. For example, one can build as faithfully as possible a neural net model of a specific portion of the brain based on the current physiological data. Then, by investigating the model by computer simulation, one tries to explain and predict the neural phenomena that might take place in that part of the central nervous system. This is referred to as the bottom-up method. The top-down approach, on the other hand, aims at synthesizing a specific function which the brain performs. Artificial intelligence is an extreme example of this method in which "intelligent behavior" is realized in computer programs with little reference to underlying neural mechanisms.

The mathematical method of approach, which the present paper is based on, intends neither to build directly a model of a specific portion of nor to synthesize a specific function of the brain. It tries, instead, to understand from the mathematical viewpoint the characteristic features and general capabilities which systems composed of neural elements naturally possess. We begin with the mathematical analysis of rather abstract neural nets whose structures might commonly be found in various forms in different parts of the brain. In order to make the mathematical analysis tractable and to obtain a clear insight of the capabilities of such neural nets, we try to simplify the models as much as possible so long as the model nets maintain the same intrinsic behaviors. We are thus able to obtain a full understanding about the behaviors of neural nets of typical structures. These typical nets must function as the building blocks in the real neural systems. Hence, this knowledge helps in constructing a specific model of a portion of the brain as well as in synthesizing brain-like information-processing systems. Also provided herein are various mathematical techniques for the analysis of neural systems.

This article is an expository paper integrating the author's recent endeavor in constructing a neural net theory (Amari, 1971, 1972a,b, 1974a,b, 1975, 1977a,b; Amari, Yoshida, and Kanatani, 1976). Without discussing the details of mathematical proofs, an attempt is made to provide a thorough comprehension of models, mathematical methods, and results.

The paper consists of three parts. The first treats dynamics of neuron pools in which connections among neurons may be random. In such a neuron pool, the behavior of an individual neuron is not important. We only need to know the behavior of a net as a whole by considering the dynamics of the macroscopic quantities. The dynamics of a simple neuron pool are monostable or bistable depending on the parameters of the net as well as the level of stimulation. This elucidates the catastrophe type of the dynamics of a neuron pool. A bistable neuron pool demonstrates hysteresis and, hence, is an important building block of information-processing systems. In this first section we will also derive the dynamical equation of systems of neuron pools. Some of these admit stable neural oscillation, while others are multistable, playing an essential role in various neural phenomena. (See, e.g., the cooperation-competition model of Amari and Arbib, this volume.)

In the second section we consider the dynamics of pattern formation in homogeneous neural fields. A neural field may be regarded as an aggregate of spatially distributed neuron pools. We shall show various types of pattern dynamics in the field (cf. the reaction-diffusion equation, Turing, 1952; Levin, 1974). A field has the ability of searching for the position of the maximum stimulus input and forming a localized excitation pattern at that point. Another field has the ability of propagating a traveling excitation pattern with shaping effect. Our method gives a mathematical foundation of the results obtained by Wilson and Cowan (1973). The theory is applied to the analysis of the cooperation-competition model (Amari and Arbib, this volume).

The third part deals with the self-organizing or learning aspects of neural nets with the intent to search for a neural mechanism common to higher-level information processing, such as association and concept-formation. For this purpose, two types of nets are considered. One is a net which transforms an input pattern signal into an output pattern signal. Given a set of input and output pattern pairs, the net is required to realize this input/output relation by self-organizing its structure. The other type of net is a sequential net having recurrent connections. The net is multistable, having a number of stable equilibrium states. We compare a stable state to a concept pattern which the net has formed. Given a number of clusters of patterns, the net is required to form a stable equilibrium pattern for each cluster by self-organizing its structure automatically. We propose two learning methods, called orthogonal and covariance learning, which realize the above self-organization. The theory is an extension of the result obtained by Amari (1972b), Anderson (1972), Kohonen (1972, 1974), Nakano (1972), Uesaka and Ozeki (1972), and Wigström (1973). It provides a common basis for the distributed and superimposed manner of keeping knowledge in a neural net.

## 2. *DYNAMICS OF NEURON POOLS*

### 2.1 *Neuron Pools*

A neuron pool in which neurons may have mutual connections can be considered as a building block of a larger-scale neural system. All the neurons in a neuron pool cooperate together to perform a role. In order to understand the role of such a neuron pool, it is not important to know the behavior of a specific neuron in the pool. It is necessary to know only the behavior of the neuron pool as a whole or, in other words, the macroscopic behavior of the neuron pool. The macroscopic behavior is represented by the dynamics of macroscopic quantities, such as the average firing rate of the neuron pool. We will treat the dynamics of neuron pools as well as the dynamics of systems of mutually interconnected neuron pools from the macroscopic point of view.

A neuron pool, which is the simplest unit carrying the macroscopic information, consists of neurons of the same type. All the neurons in a neuron pool have the same connection parameters except for random fluctuations. The effect of this randomness is also taken into account, which leads us to statistical neurodynamics (Amari, 1974a; Amari et al., 1976).

Consider a simple model of a neuron element receiving n inputs $y_1, \ldots, y_n$ and emitting one output x (Figure 1). Let $w_1, \ldots, w_n$ be the so-called synaptic weights of n inputs. We call the weighted sum of the inputs

$$u = \sum_{i=1}^{n} w_i y_i$$

the potential. The output is determined by a function of u,

$$x = f(u - h), \tag{2.1}$$

where h is the threshold value.

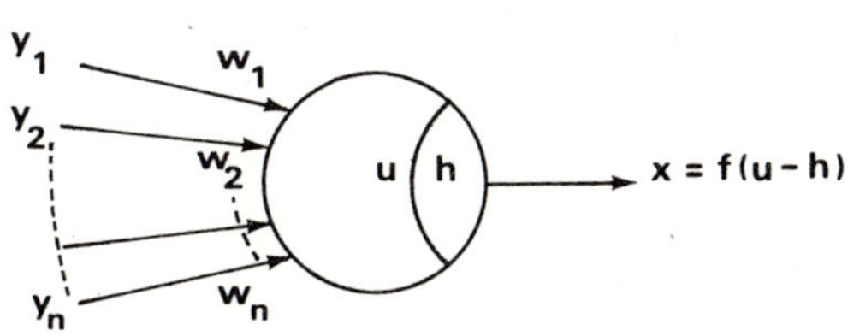


*Figure 1* Neuron model.

The simplest model is the McCulloch-Pitts formal neuron (McCulloch and Pitts, 1943) in which the input and output variables take on two values, 0 or 1, corresponding to the existence or nonexistence of an excitation pulse. In this case, the function f has the form

$$f(u) = \begin{cases} 1 & u > 0, \\ 0 & u \le 0. \end{cases} \tag{2.2}$$

We can consider a more realistic model in which $y_i$'s and x take on real values representing the respective firing rates. The function f in this case is a monotonically increasing, continuous function having a saturation character. We may treat a continuous-time version of the model in which the potential u(t) is subject to the following differential equation:

$$\begin{cases} \tau \dfrac{du(t)}{dt} = -u(t) + \Sigma w_i y_i(t), \\ x(t) = f[u(t) - h], \end{cases}$$

where t denotes the continuous-time variable.

It is interesting to note that any of the above models give qualitatively the same result as far as the macroscopic dynamics of a neuron pool are concerned (Amari, 1971, 1972a, 1974b). We will, therefore, use the McCulloch-Pitts model for simplicity's sake.

### 2.2 *Neuron Pool Without Recurrent Connections*

Let us consider a neuron pool consisting of m elements (Figure 2)

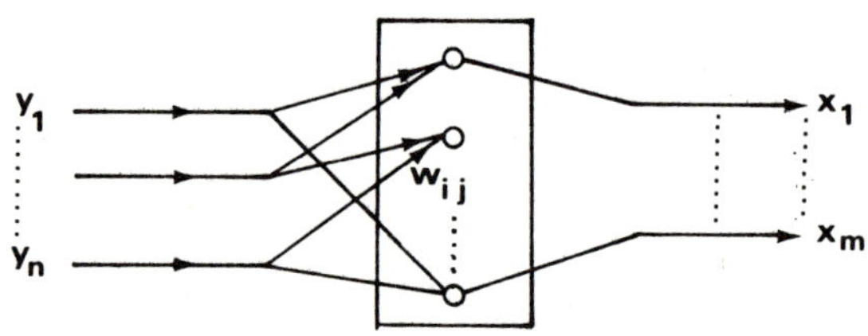


*Figure 2* Neuron pool without recurrent connections.

and let $x_i$ be the output of the *i*th element. The net receives m inputs $y_1,\ldots,y_m$. Let $w_{ij}$ be the synaptic weight of $y_j$ connected to the *i*th element. Let $h_i$ be the threshold of the *i*th element. The output $x_i$ is then written as

$$x_i = f\left(\sum_{j=1}^{n} w_{ij} y_j - h_i\right), \qquad i = 1,\ldots,m.$$

We now study the input/output relation of the pool from the macroscopic point of view. We define the activity of the input signals by

$$s = \frac{1}{n} \sum_{i=1}^{n} y_i \tag{2.3}$$

and the activity of the outputs $x_1, \ldots, x_m$ of the neuron pool by

$$s' = \frac{1}{m} \sum_{i=1}^{n} x_i . \tag{2.4}$$

The problem is to determine the relation between the input activity s and the output activity s'. The synaptic weights $w_{ij}$, $i = 1, \ldots, m$; $j = 1, \ldots, n$, are not the same. We assume that they are randomly distributed with $\bar{w}$ and $\sigma_w^2$, the average and the variance, respectively. Consider, for example, the case where an input line has a connection to a neuron with probability p and a synaptic weight of w. Then $w_{ij} = w$ if the *j*th input is connected with the *i*th element and $w_{ij} = 0$ otherwise. In this case, the average of $w_{ij}$ is

$$\bar{w} = pw$$

and the variance is

$$\sigma_w^2 = w^2 p(1 - p).$$

We also assume that the thresholds $h_i$'s are normally distributed with the average value $\bar{h}$ and the variance $\sigma_h^2$.

Given inputs $y_1, \ldots, y_n$, the potential $u_i$ of the *i*th element becomes

$$u_i = \Sigma\, w_{ij} y_j .$$

Since $w_{ij}$'s and $h_i$ are different for different i, $(u_i - h_i)$'s are also different for different i. The expected value of $(u_i - h_i)$ is given by

$$\bar{u} = \overline{(\Sigma\, w_{ij} y_j - h_i)} = \bar{w}\, \Sigma\, y_j - \bar{h} = n\bar{w}s - \bar{h},$$

while the variance of $(u_i - h_i)$'s is given by

$$\sigma^2 = n\sigma_w^2 s + \sigma_h^2,$$

where we take into account that $y_j$'s assume two values, 0 and 1. When n is large, we can assume by virtue of the central limit theorem that $(u_i - h_i)$'s are normally distributed with an expected value $(\bar{u} - \bar{h})$ and variance $\sigma^2$.

An element is excited when it has positive $(u_i - h_i)$. Therefore, among n elements, those which have positive $(u_i - h_i)$ are excited. The ratio of excited elements is given by the probability that $(u_i - h_i)$ becomes positive, or

$$\begin{aligned}\Phi(s) &= \int_0^{\infty} \frac{1}{\sqrt{2\pi}\sigma} \exp\left\{\frac{-(z - \bar{u})^2}{2\sigma^2}\right\} dz \\ &= \Psi\left(\frac{n\bar{w}s - \bar{h}}{\sqrt{n\sigma_w^2 s + \sigma_h^2}}\right)\end{aligned} \tag{2.5}$$

where $\Psi(x)$ is the integration of the normal distribution

$$\Psi(x) = \int_{-\infty}^{x} \frac{1}{\sqrt{2\pi}} \exp\left\{-\frac{v^2}{2}\right\} dv. \tag{2.6}$$

The rate of the excited elements is the activity of the output. Hence, we have the macroscopic input/output relation

$$s' = \Phi(s). \tag{2.7}$$

We have thus obtained the input/output relation of a neuron pool for the simplest model. When the function f is more general and $y_j$'s and $x_i$'s take on real values, we can obtain the input/output relation in a similar manner (Amari, 1972b, 1974a; Amari et al., 1976).

There is a theoretical problem with the assumption that $u_i$'s are normally distributed or, more generally, a problem concerning stochastic structures of a neuron pool. We do not consider this problem here (see, however, Amari et al., 1976). Nevertheless, the result derived from this simple model can explain very well the macroscopic behavior of neuron pools.

*2.3 Neuron Pool With Recurrent Connections*

We now consider a neuron pool in which the output of a neuron is connected with other neurons. The output signals, in this case, are fed back to the input of the net (Figure 3). Therefore, the

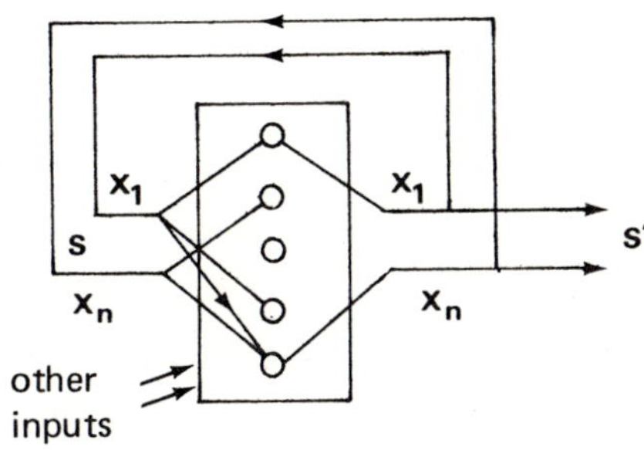


*Figure 3* Neuron pool with recurrent connections.

present activity of the pool

$$s = \frac{1}{n} \sum_{i=1}^{n} x_i$$

yields the subsequent input activity of the pool. Thus, the subsequent output is determined by

$$s' = \Phi(s).$$

Assuming discrete times $t = 1,2,3,\ldots$, we have the dynamical equation

$$s_{t+1} = \Phi(s_t), \qquad t = 1,2,\ldots, \tag{2.8}*$$

*On deriving this equation rigorously, we encounter a mathematical difficulty similar to that which we encounter in statistical mechanics. Part of the difficulty is solved in Amari et al. (1976). See, also, Rozonoer (1969a,b,c) and Amari (1974a).

where $s_t$ is the activity of the neuron pool at time t. The dynamical properties of the neuron pool are revealed by this equation which includes the parameters $\bar{w}$, $\bar{h}$, $\sigma_w^2$, and $\sigma_h^2$. When additional input stimuli arrive at other input lines of this neuron pool from the outside, activating the neurons, the threshold values of the neurons decrease by these amounts of stimulation. Hence, the stimulation results in a decrease in $\bar{h}$ of the neuron pool. Accordingly, the parameter $\bar{h}$ of the neuron pool can be controlled through the input stimulation from the outside. ($\sigma_h^2$ may also be changed by the stimulation.)

An activity $\tilde{s}$ which satisfies

$$\tilde{s} = \Phi(\tilde{s}) \tag{2.9}$$

is called an equilibrium. A neuron pool is said to be *monostable* when it has one and only one equilibrium; moreover, the activity eventually converges to that equilibrium irrespective of the initial level of activity. A neuron pool is said to be *bistable* when it has three equilibria, $\tilde{s}_1$, $\tilde{s}_2$, $\tilde{s}_3$, with $\tilde{s}_1 < \tilde{s}_2 < \tilde{s}_3$, $\tilde{s}_1$ and $\tilde{s}_3$ being stable and $\tilde{s}_2$ being unstable, and the activity converging to $\tilde{s}_1$ or $\tilde{s}_3$ depending on whether the initial activity is smaller or larger than $\tilde{s}_2$. A neuron pool is said to be oscillatory when the activity must always fall into an oscillation.

We can show that a neuron pool is monostable, bistable, or oscillatory, depending on the parameters $\bar{w}$, $\bar{h}$, $\sigma_w^2$, and $\sigma_h^2$. No other types of dynamics exist as far as a neuron pool of this simple kind is concerned. The dynamics of a simple neuron pool have been studied by many investigators (e.g., Allanson, 1956; Ashby, Foerster, and Walker, 1962; Smith and Davidson, 1962; Annios, Beek, Csermely, and Harth, 1970). Its parameter dependency is elucidated theoretically by Amari (1971, 1972a). When $\bar{w}$ is negative and its absolute value is sufficiently large, a neuron pool becomes oscillatory, its activity alternating a high value and a low value with period 2. We have little interest in this period 2 oscillation because it is derived by parsing time too roughly. If we treat a continuous-time version of the model, such oscillation does not occur, and we have monostable and bistable neuron pools (see Amari, 1972a). A neuron pool consisting of neuron elements with an absolute refractory period also has the same characteristics (Yoshizawa, 1974).

When $\bar{w}$ is small, the neuron pool is monostable for any value of $\bar{h}$. Since the average threshold $\bar{h}$ changes with the intensity of the additional extrinsic sources of stimulation, we find that

$$\bar{h} = \bar{h}_0 - I,$$

where I denotes the intensity of the additional stimulation. The equilibrium activity $\tilde{s}$ of a monostable neuron pool is uniquely determined, depending on $\bar{h}$ or I. Thus,

$$\tilde{s} = \tilde{s}(I).$$

This is demonstrated in Figure 4. The equilibrium activity naturally increases as I increases.

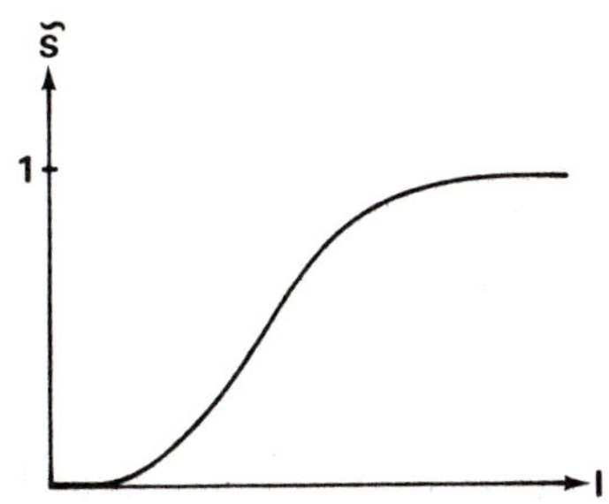


*Figure 4* Equilibrium activity as a function of input stimulation for a monostable neuron pool.

When $\bar{w}$ is large, a neuron pool is bistable for an adequate value of $\bar{h}$ or I. When I becomes too big or too small, the pool again becomes monostable. The equilibrium activities of a bistable neuron pool are determined by the additional extrinsic stimulation I as

$$\tilde{s} = \tilde{s}(I).$$

This is depicted in Figure 5 where the net is bistable if I is in the interval

$$I_1 < I < I_2.$$

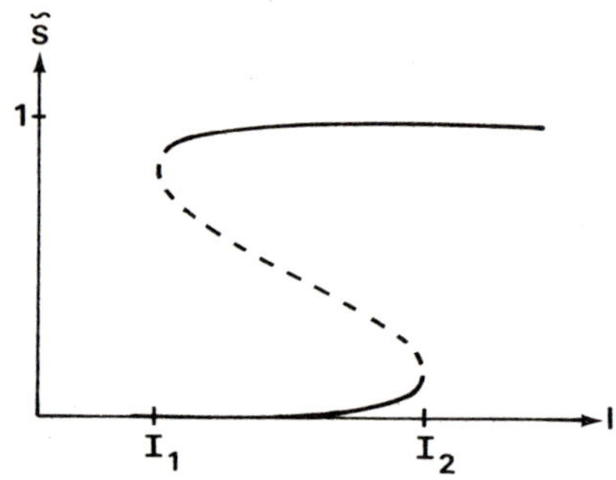


*Figure 5* Equilibrium activity as a function of input stimulation for a bistable neuron pool.

Within this interval, the function $\tilde{s}$ has three values; the high and low values in the solid lines depict stable equilibria while the middle dotted line denotes the unstable equilibrium.

As I changes, the equilibrium activity of the net changes, demonstrating hysteresis. For example, when I is small, the net is monostable having a low activity level $\tilde{s}$. As I increases, the activity increases very slowly but still remains at the lower equilibrium level despite the fact that the higher level becomes stable. However, when I becomes bigger than $I_2$, the activity jumps to the upper equilibrium value. Once the activity of the net has reached the upper equilibrium state, the high activity level is maintained even when I becomes smaller than $I_2$. The net returns to the low activity only when I becomes smaller than $I_1$, $I_1 < I_2$.

A bistable neuron pool is an important building block of neural networks as it provides the system with multistability and hysteresis. Hysteresis and multistability are characteristic of information processing in the brain. We can approximate the behavior of a bistable neuron net by

$$\tilde{s}' = f[(I_2 - I_1)\,\tilde{s} + I - I_2], \tag{2.10}$$

where f is the two-valued function defined by (2.2). In this simplification, $\tilde{s}$ takes only two values, 1 for the high activity level and 0 for the low activity level. This behavior can be realized by a simple threshold element as shown in Figure 6. The

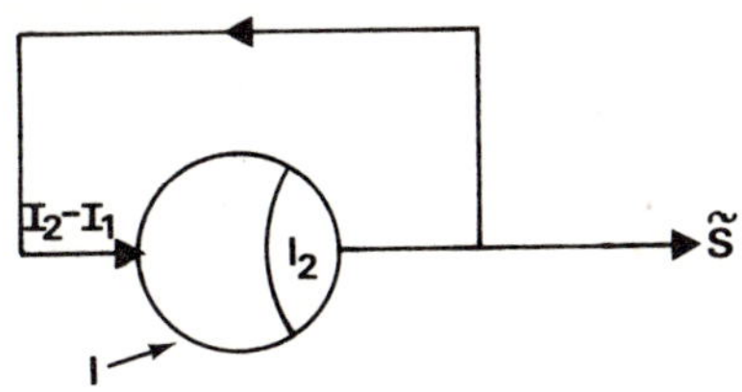


*Figure 6* A simplified model of a bistable neuron pool.

feedback input to this element has a connection weight $I_2 - I_1$ with a threshold value of $I_2$. The characteristics of this are shown in Figure 7. Such simplification is sometimes convenient for building complex models. (See, e.g., Amari, 1972b; Amari and Arbib, this volume.)

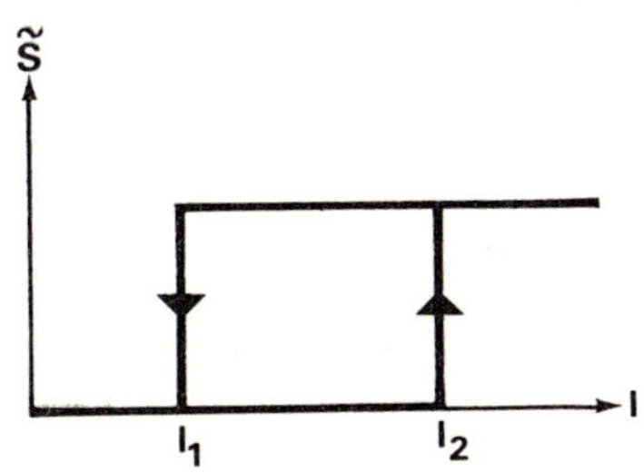


*Figure 7* Equilibrium of the simplified bistable model.

The essential feature of the dynamics of a simple neuron pool is its monostable and bistable character with its dependency on the network parameters. The dependency of the stability on these parameters is clearly shown by writing the values $\tilde{s}$ of the equilibria for two parameters, $\bar{w}$ and I (or $\bar{h}$). In the bistable region (Figure 8b), $\tilde{s}$ has three values, $\tilde{s}_1$, $\tilde{s}_2$, $\tilde{s}_3$ (Figure 8a). This

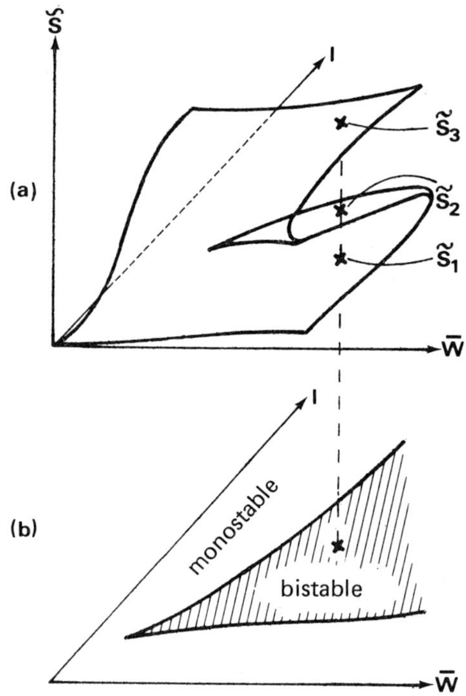


*Figure 8* Catastrophe of dynamics of a simple neuron pool.

type of parameter dependency of dynamics is called the cusp catastrophe (Thom, 1972). The catastrophe curve of the dynamics of neuron pools has been studied in detail by Amari (1971). When we adopt more plausible models of neurons, such as analog neurons obeying differential equations (Amari, 1972a) or neurons having a refractory period (Yoshizawa, 1974), the dynamics of a neuron pool are still monostable or bistable, demonstrating the same type of catastrophe. Hence, we may say that the essential feature of the dynamics of a simple neuron pool is described by the cusp catastrophe.

### 2.4 *Systems of Neuron Pools*

Consider a system composed of a number of mutually connected neuron pools (Figure 9). Let m denote the number of neuron pools, $s_i$ the activity of the *i*th neuron pool, and $\bar{w}_{ij}$ and $\sigma_{ij}^2$ the average and the variance, respectively, of the synaptic weights of connections from the neurons of the *j*th neuron pool to those of the *i*th neuron pool. A neuron pool may also have recurrent connections within itself. If it has no recurrent connections, $w_{ij}$ is equal to 0. In addition, let $n_i$ represent the number of elements in the *i*th pool and $\bar{h}_i$ and $\sigma_{hi}^2$ the average and the variance, respectively, of the thresholds of the neurons in the *i*th neuron pool.

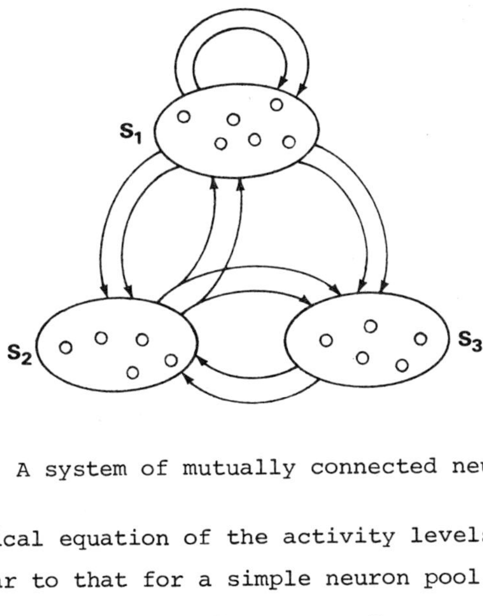


*Figure 9* A system of mutually connected neuron pools.

The dynamical equation of the activity levels is obtained by a method similar to that for a simple neuron pool; i.e.,

$$s_i(t + 1) = \Psi \left( \frac{\Sigma\, n_j \bar{w}_{ij} s_j(t) - \bar{h}_i}{\sigma_i} \right), \qquad (2.11)$$

$$i = 1, \ldots, m,$$

where

$$\sigma_i^2 = \sum_{j=1}^{n} n_j \sigma_{ij}^2 s_j(t + 1) + \sigma_{hi}^2.$$

The system is, in general, multistable. However, an oscillatory solution of a large period is possible when the parameters are adequately chosen. When the variances $\sigma_{ij}^2$ are small, we can ignore these terms and thereby obtain the following simplified equation:

$$s_j(t+1) = \Psi[\Sigma\, w_{ij} s_j(t) - h_i], \tag{2.12}$$

where

$$w_{ij} = \frac{n_j \bar{w}_{ij}}{\sigma_i},$$

$$h_i = \frac{\bar{h}_i}{\sigma_i}.$$

If we approximate the function $\Psi$ by f of (2.2), we get

$$s_j(t+1) = f[\Sigma\, w_{ij} s_j(t) - h_i], \tag{2.13}$$

$s_j$ taking on the values 0 and 1. The simplified equation corresponds to the simplification (2.10). It still represents the multistability and the oscillatory character of the system very well. A mathematical analysis of such systems has been provided by Amari (1972b).

When we adopt the continuous-time model, the equation of a system of neuron pools is derived as

$$\begin{cases} \dfrac{du_i}{dt} = -\,u_i + \Sigma\, w_{ij} x_j - h_i, \\ x_i = \varphi(u_i), \end{cases} \tag{2.14}$$

where $u_i$ and $x_i$ denote the average membrane potential and the activity (i.e., average firing rate), respectively, of the *i*th neuron pool (Amari, 1972a).

### 2.5 *Neural Oscillations*

Systems of neuron pools may be characterized by oscillatory behaviors. Amari (1971, 1972a) has shown that a system composed of both an excitatory and an inhibitory neuron pool has a persistent oscillation and has described the oscillatory conditions in detail. A similar result has also been demonstrated by Wilson and Cowan (1972) using the same configuration of neuron pools. In these models, the excitatory neuron pool must have recurrent connections (Figure 10a).

A system consisting of two excitatory neuron pools and one inhibitory neuron pool also has a persistent oscillation. In this system, no direct self-recurrent connections are necessary (Figure 10b).

Lieblich and Amari (this volume) have proposed the neuron-pool system of Figure 10c to explain kindling phenomena. It has a stable oscillatory orbit around a stable equilibrium state. Ahn and Freeman (1974) have proven the existence of a system of four neuron pools, which is equivalent to a system consisting of second-order dynamics.

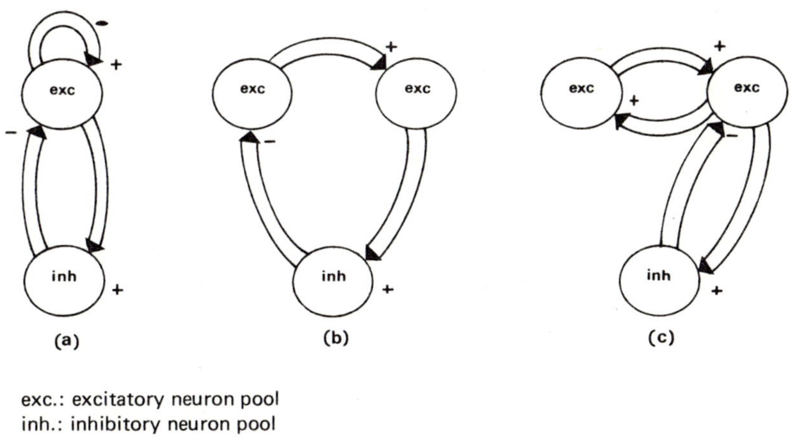


*Figure 10* Oscillatory systems.

## 3. *DYNAMICS OF PATTERN FORMATION IN NEURAL FIELDS*

### 3.1 *Dynamical Equation of Neural Fields*

We now consider the dynamics of excitation patterns in a neuron field consisting of m layers. The neurons in a neuron field may have random connections within layers and between layers. Let $u_i(x,t)$ represent the potential at position x and time t of the *i*th layers, and $w_{ij}(x,y)$ the average of the synaptic weights of connections from the neurons at position y of the *j*th layer to the neurons at position x of the *i*th layer. Further, let $h_i$ denote the average of the thresholds of the neurons of the *i*th layer, and $I_i(x,t)$ the input stimulus arriving at position x at time t of the *i*th layer.

We can then develop the continuous-time version of the equation of a neuron pool system (2.14) into the field equation

$$\tau_i \frac{\partial u_i(x,t)}{\partial t} = -u_i + \sum_{j=1}^{m} \int w_{ij}(x,x')X_j(x',t)dx' - h_i + I_i(x,t), \tag{3.1}$$

$$X_i(x,t) = f_i[u_i(x,t)],$$

where $\tau_i$ is the time constant of the dynamics of the neurons of the *i*th layer; $X_i(x,t)$ is the output activity of the neurons at position x, time t of the *i*th layer; and $f_i$ is the function connecting the potential with the output activity. $W_{ij}$ and $h_i$ are normalized by dividing by $\sigma$. The equation is similar to that derived by Wilson and Cowan (1973).

We are interested in the multistable character of a neural field* which gives rise to hysteretic phenomena, pattern formation, and neural oscillation. As far as the phenomena related to multistability are concerned, we can consider the simplified equation in which the output function $f_i$ is the step function f defined in (2.2). Given this simplification, each layer can be divided into two regions; namely, an excitatory region in which f(u)

*There are many interesting mathematical analyses of neural fields (see, e.g., Reichardt and MacGinite, 1962; Hadler, 1974; Oguztöreli, 1975). However, they have only considered monostable cases. Ellias and Grossberg (1975), on the other hand, have recently provided a mathematical analysis capable of taking into account multistable cases.

is 1 and an unexcitatory region in which f(u) is 0. By noting the growth or contraction of the excitatory region, we can study the dynamics of field excitation. This gives us an analytical method of solving the field equation explicitly. Results obtained from this simplified equation can explain qualitatively the behavior of more plausible field equations. They can, for example, account for all the phenomena of the Wilson-Cowan model (Wilson and Cowan, 1973).

### *3.2 Pattern Formation in Lateral-Inhibition Fields*

Consider a homogeneous field of one layer whose connections are of a lateral-inhibitory type; i.e., excitatory connections dominate among the proximate neurons, while inhibitory connections are formed with the more distal neurons.

By virtue of homogeneity, the connection weight between a neuron at x and one at y depends only on their difference x - y. Thus,

$$w_{ij}(x,y) = w(x - y).$$

We assume that the field is one-dimensional merely for simplicity's sake. In addition, we assume the isotropy of the field,

$$w(x) = w(-x). \tag{3.2}$$

The typical shape of w(x) of a lateral-inhibitory field is shown in Figure 11. In this case, the field equation is

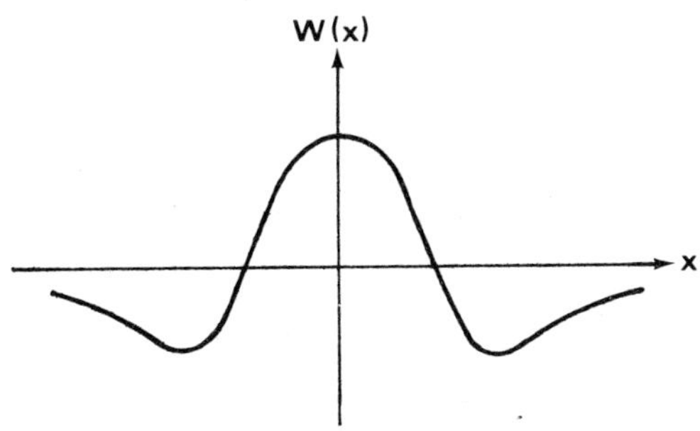


*Figure 11* Weighting function of a lateral-inhibitory field.

$$\tau \frac{\partial u(x,t)}{\partial t} = - u(x,t) + \int w(x - y) f[u(y,t)] dy - h + I. \tag{3.3}$$

We will now demonstrate the formation of excitation patterns in a homogeneous field.* Our discussion does not rely on a specific kind of connection so long as it is of the lateral-inhibitory type. We first consider the case where input I is spatially uniform. We then study the effect of nonuniform input.

Assume that excitation is concentrated in the region from $x_1$ to $x_2$ at time t, the potential function u(x,t) being positive only in the interval $(x_1,x_2)$ (Figure 12). The region of excitation will change with time according to the field dynamics (3.3).

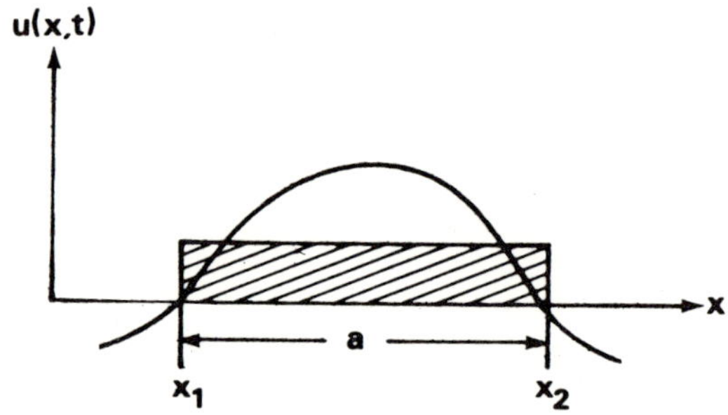


*Figure 12* A localized region of excitation.

If the excitatory region moves to $(x_1 + dx_1,\ x_2 + dx_2)$ after time dt, the boundaries of the region will change by $dx_1$ and $dx_2$, respectively. We then have

$$u(x_i + dx_i,\ t + dt) = 0, \qquad i = 1,2,$$

since $(x_i + dx_i)$ is the boundary at time t + dt. Taylor expansion

*We have reaction-diffusion systems as a model of pattern formation in a homogeneous field (see, e.g., Turing, 1952; Bard and Lauder, 1974; Maginu, 1975). The present model provides a simpler mechanism of pattern formation.

yields

$$\frac{\partial u(x_i,t)}{\partial x}\,dx_i + \frac{\partial u(x_i,t)}{\partial t}\,dt = 0,$$

from which we obtain

$$\frac{dx_i}{dt} = -\frac{\partial u(x_i,t)}{\partial t}\bigg/\frac{\partial u(x_i,t)}{\partial x}.$$

This represents the velocity of motion of the boundary points.

Let us put

$$c_1 = \frac{\partial u(x_1,t)}{\partial x}$$

and

$$c_2 = -\frac{\partial u(x_2,t)}{\partial x},$$

the gradients of the waveform $u(x,t)$ at the boundary points, with $c_1, c_2 > 0$. From (3.3) we have

$$\frac{\partial u(x_i,t)}{\partial t} = \int_{x_1}^{x_2} w(x_i - y)\,dy - h + I$$

since

$$u(x_i,t) = 0, \qquad i = 1,2,$$

and

$$f[u(x,t)] = 1$$

in the interval $(x_1,x_2)$.

Let $W(x)$ be the integration of the connection function

$$\int_0^x w(y)\,dy = W(x). \tag{3.4}$$

Since

$$\int_{x_1}^{x_2} w(x_i - y)\,dy = W(x_2 - x_1), \qquad i = 1,2,$$

we arrive at the equations governing the change in the region of excitation,

$$\frac{dx_1}{dt} = -\frac{1}{\tau c_1}[W(x_2 - x_1) - h + I]$$

and

$$\frac{dx_2}{dt} = \frac{1}{\tau c_2}[W(x_2 - x_1) - h + I]. \tag{3.5}$$

The length of the excitatory region is shown by

$$a = x_2 - x_1,$$

while the change in length is represented by

$$\frac{da}{dt} = \frac{dx_2}{dt} - \frac{dx_1}{dt}$$

$$= \left(\frac{1}{\tau c_2} + \frac{1}{\tau c_1}\right)[W(a) - h + I]. \tag{3.6}$$

We have thus succeeded in reducing a complex, partial integro-differential equation of the field (3.3) to a simple, ordinary differential equation of the length of the excitatory region (3.6). When

$$\frac{da}{dt} < 0$$

or

$$W(a) < h - I,$$

the excitatory region shrinks. On the contrary, when

$$W(a) > h - I,$$

the excitatory region expands. Equilibrium occurs at the

length a satisfying

$$W(a) = h - I. \tag{3.7}$$

We now need to know the shape of the function $W(x)$. As is shown in Figure 13, $W(x)$ first increases until it attains its maximum value $W_m$ at the point satisfying $w(x) = 0$ and then decreases, converging to $W_\infty$. When $W_\infty > 0$, the excitatory connections as a whole are stronger than the inhibitory ones. When $W_\infty < 0$, the inhibitory connections dominate.

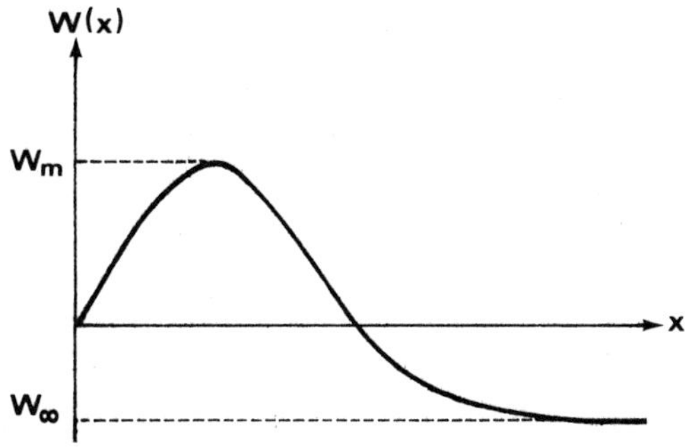


*Figure 13* The shape of $W(x)$.

Dynamical properties of pattern formation in a homogeneous field are revealed by (3.7) and Figure 13 in the following manner. Let us first consider an inhibition-dominated field ($W_\infty < 0$).

i. As long as the intensity of the uniform stimulation I is weaker than $h - W_m$, i.e.,

$$I < h - W_m,$$

then

$$\frac{da}{dt} < 0.$$

Consequently, any excitatory region will shrink until it eventually disappears. The field cannot maintain any excitatory activity.

ii. As I increases, satisfying

$$h > I > h - W_m,$$

equation (3.7) has two solutions, $a_0$ and $a_1$, with $a_0 < a_1$ (Figure 14). Localized excitations of length $a_0$ and $a_1$ are, therefore,

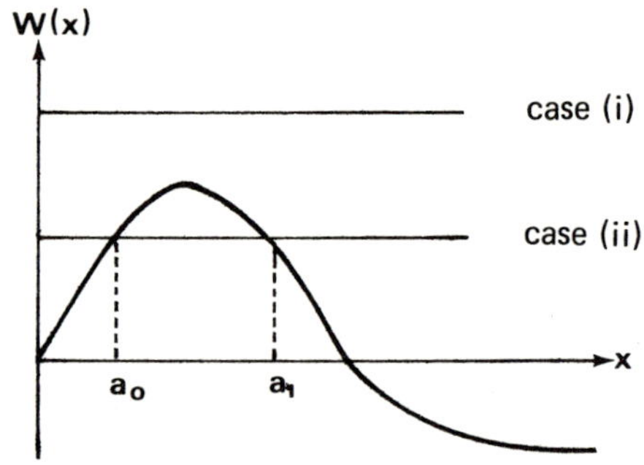


*Figure 14* Equilibrium excitation length.

the equilibrium solutions of the field. However, the solution with length $a_0$ is unstable because when $a > a_0$,

$$\frac{da}{dt} > 0,$$

but when $a < a_0$,

$$\frac{da}{dt} < 0.$$

This means that whenever the region of excitation becomes a little larger than $a_0$, it expands further, but whenever it becomes a little smaller than $a_0$, it shrinks to nil. The solution with length $a_1$ is, on the contrary, stable. Hence, the field can maintain a localized pattern of excitation of length $a_1$. The field is, therefore, multistable. Whenever the initial excitation is smaller than $a_0$ in length, the excitation dies out, whereas whenever it is larger than $a_0$, the excitation is maintained, being shaped to length $a_1$.

iii. As I becomes much larger, satisfying

$$h < I < h - W_\infty,$$

equation (3.7) has only one solution, $a_1$. Obviously,

$$\frac{da}{dt} > 0 \qquad \text{when } a < a_1$$

and

$$\frac{da}{dt} < 0 \qquad \text{when } a > a_1.$$

Therefore, the quiescent state, in which no part of the field is excited, is not permitted in this case. The field always has a spontaneous excitation of length $a_1$.

iv. When the stimulation I is so strong that

$$I > h - W_\infty,$$

$$\frac{da}{dt} > 0$$

holds for all a. Hence, any single excitation pattern grows, apparently without limit, provided no long-range interaction with other excited patterns exists. When $I > h - 2W_\infty$, the entire field becomes highly excited. However, when $I < h - 2W_\infty$, by virtue of the mutual interaction of excited patterns, the entire field is not excited. Instead, a series of intermittent excitation patterns is aroused.

The results are similar for a field dominated by excitatory connections ($W_\infty > 0$).

i. When

$$I < h - W_m,$$

the excitation dies out completely.

ii. When

$$h - W_m < I < h - W_\infty,$$

equation (3.7) has two solutions, $a_0$ and $a_1$. The field is thus multistable: the excitation pattern of length $a_1$ is stable while that of length $a_0$ is unstable.

iii. When

$$h - W_\infty < I < h,$$

equation (3.7) has only one solution, $a_0$. However, an excitation pattern of length $a_0$ is unstable. Hence, if the initial excitation is smaller than $a_0$, it dies out; if it is larger than $a_0$, it spreads over the entire field.

iv. When

$$I > h,$$

the entire field becomes excited.

A localized excitation pattern, when it exists, can occur at any place in the field since the field is homogeneous. However, when the input stimulation I(x) is nonuniform, depending on position x, the localized excitation moves in the direction of increasing stimulation in search of the position of the maximum stimulation. We can demonstrate this as follows: When an additional term I(x) is present, the equations for the motion of the boundary points are modified as

$$\frac{dx_1}{dt} = -\frac{1}{\tau c_1}[W(a) - h + I + I(x_1)]$$

and

$$\frac{dx_2}{dt} = \frac{1}{\tau c_2}[W(a) - h + I + I(x_2)].$$

By combining the above two equations and letting $c_1 = c_2 = c$,*

$$\frac{1}{2}\frac{d(x_1 + x_2)}{dt} = \frac{1}{2\tau c}[I(x_2) - I(x_1)].$$

*This assumption is not necessary. However, it greatly simplifies the following discussion.

Since $\frac{1}{2}$ $[d(x_1 + x_2)/dt]$ denotes the velocity of motion of the center of the excitatory region, we observe that the region of excitation moves to the right if $I(x_2) > I(x_1)$ and to the left if $I(x_2) < I(x_1)$, i.e., in the direction of increasing stimulation, stopping when $dx_1/dt = dx_2/dt = 0$. It occurs at the position where

$$I(x_1) = I(x_2).$$

Consequently, the excitatory region eventually locates the maximum of $I(x)$ and extends over the peak.

Thus far we have considered only a solitary, localized pattern of excitation. We also need to determine the interaction among these localized excitation patterns. When a localized pattern of excitation of length a exists in the interval $(-a,0)$, the neurons in that interval send output stimulation to neurons of other positions. The total amount of stimulation received by the neurons located at place x $(x > 0)$ is

$$S(x) = \int_{-a}^{0} w(x - y)dy = W(x + a) - W(x). \tag{3.8}$$

The function $S(x)$ generally has a shape shown in Figure 15. Hence, a localized excitation pattern is the source of stimulation $S(x)$ for another part of the field. If another localized pattern

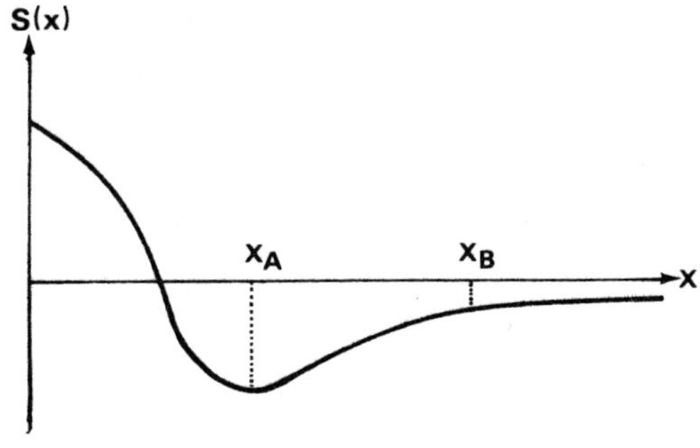


*Figure 15* Interacting stimulation from a localized excitation.

of excitation exists in the field, this second pattern moves in the direction of increasing S(x) which has been aroused by the initial excitation, and vice versa. Therefore, when two localized excitations exist within a distance of $x_A$, the minimum of S(x), they are mutually attracted and eventually unite, forming a single excitatory region; when two localized excitations exist at a distance greater than $x_A$, they repel each other and move in opposite directions until no interactions exist. No interactions exist beyond the distance $x_B$ so that any patterns may coexist beyond this distance.

This type of simple field can be utilized as a building block of a complex system. For example, it can function as a short-term memory of persisting excitation or as a pattern generator; or it can serve to locate a maximum stimulus or to encode the intensity of a stimulus into the length of the excited region. This theory has been applied to the analysis of a cooperation and competition model (Amari and Arbib, this volume).

### *3.3 Dynamic Pattern Formation in a Field of Excitatory and Inhibitory Layers*

In order to understand dynamic pattern formations, such as oscillatory and traveling wave patterns, we need first to consider the interaction between the excitatory and inhibitory layers of a field. Let a field consist of a layer of excitatory neurons and a layer of inhibitory neurons. Let $w_1(x)$ and $w_2(x)$ represent the weight functions of the connections from neurons in the inhibitory layer and excitatory layer, respectively, to neurons in the excitatory layer. We also assume, for simplicity's sake, that there are no direct connections among inhibitory neurons and that an inhibitory neuron at x receives stimulation only from the excitatory neurons located at the same position x in the field. The synaptic weight of these connections is $w_3$.

We then have the following field equation:

$$\tau \frac{u_1(x,t)}{\partial t} = - u_1 + \int w_1(x - y) f[u_1(y,t)]dy - \int w_2(x - y) f[u_2(y,t)]dy - h_1 + I(x,t), \tag{3.9}$$

$$\tau \frac{\partial u_2(x,t)}{\partial t} = - u_2 + w_3 f[u_1(x,t)] - h_2,$$

where $u_1(x,t)$ and $u_2(x,t)$ denote the potentials of the excitatory and inhibitory layers, respectively, at position x and time t.

We first obtain a homogeneous solution of the field. Integrating (3.9) over x, we have

$$\tau \frac{du_1}{dt} = - u_1 + w_1 f(u_1) - w_2 f(u_2) - h_1 + I,$$
$$\tau \frac{du_2}{dt} = - u_2 + w_3 f(u_1) - h_2, \tag{3.10}$$

where

$$I = \int I(x,t)dx,$$
$$u_i(t) = \int u_i(x,t)dx,$$
$$w_i = \int w_i(x)dx, \qquad i = 1,2.$$

This may be regarded as the equation of a system of excitatory and inhibitory neuron pools.

It is easy to study the behavior of this system (see, e.g., Amari, 1977a). When $w_1 \geq w_2$ (i.e., when the self-excitatory connections dominate the inhibitory connections), the behavior of the system is monostable or bistable depending on the value of input stimulation I. When $w_2 > w_1$ (i.e., when the inhibitory connections are stronger than the excitatory connections), the behavior of the system is monostable, bistable, or oscillatory depending on I. The frequency of oscillation also depends on I. In the monostable case, the system exhibits the active transient (Wilson and Cowan, 1973). That is, when an input stimulus is applied to the system, the potential $u_1$ of the excitatory neuron pool continues to increase for a short time even after the cessation of the stimulus, provided the intensity and time duration of the stimulus are both larger than some threshold value. Thereafter, the potential returns to the resting value. We can also observe the dual process of the active transient.

When a part of the field is stimulated, that region first becomes excited, demonstrating the active transient, and then returns to the resting state. If the stimulation continues for a time, the activity in that part of the field may oscillate during that period. The excitation aroused may also propagate throughout the entire field as a traveling wave. We can prove the existence of such a traveling wave by extending the previous method. We can also study the behavior of the nonlinear, partial integro-differential equation (3.9) by the use of an ordinary differential equation of the boundary motion of the excitatory regions. The criterion for the existence of a stable traveling wave has been explicitly described by Amari (1977a).

For example, a field with the structure

$$w_i(x) = \frac{A_1}{2\pi\sigma_i} \exp\left\{ - \frac{x^2}{2\sigma_i^2} \right\},$$

where

$$A_1 = 2.0,$$
$$A_2 = 4.0,$$
$$\sigma_1 = 1.0,$$
$$\sigma_2 = 1.5,$$
$$w_3 = 2.0,$$
$$h_1 = -0.1,$$
$$h_2 = -1.0, \text{ and}$$
$$\tau = 1.0,$$

produces a traveling wave of velocity

$$v = 7.3.$$

The waveform is shown in Figure 16. Fields of this type might function in the brain as an oscillator, a wave generator, a controller for synergetic movement, etc.

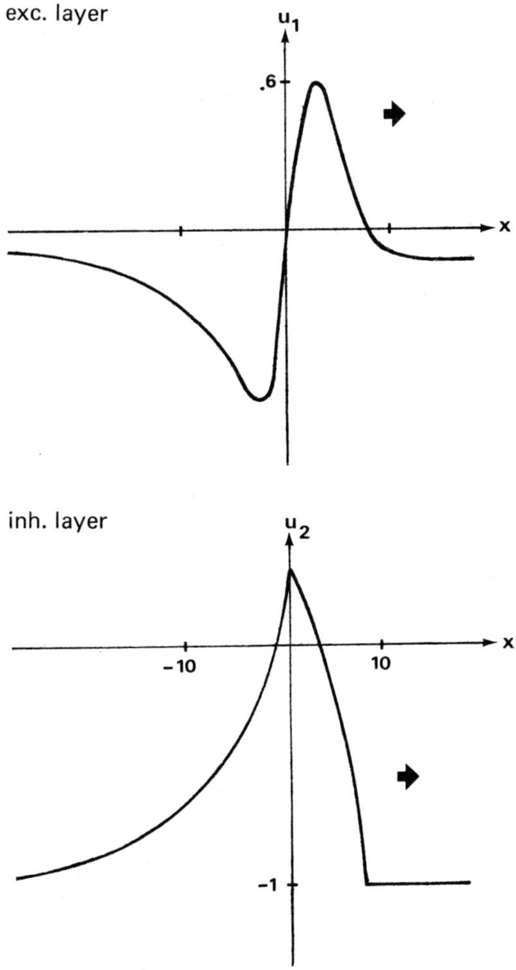


exc. layer: excitatory layer
inh. layer: inhibitory layer

*Figure 16* The shape of a traveling wave.

## 4. *A NEURAL THEORY OF ASSOCIATION AND CONCEPT FORMATION*

### 4.1 *Association, Pattern Generation, and Concept Formation*

Association, pattern generation, and concept formation are relatively high-level forms of information processing realized by neural nets with learning or self-organizing ability. With the help of primitive models, we can search for possible neural mechanisms which realize these kinds of higher-level processes. We will show how a common mechanism works well for these apparently different higher-level activities. Orthogonal and covariance learning are proposed as the common learning mechanisms, and their performance is analyzed.

A neural net which transforms an input pattern signal $x$ into an output pattern signal $y$ serves as a primitive model for association. A pattern signal is represented by a vector

$$x = \begin{bmatrix} x_1 \\ \vdots \\ x_n \end{bmatrix},$$

where the components $x_i$ represent the features of the pattern. We find that

$$y = Tx$$

when the net, receiving input $x$, outputs $y$. The transformation T depends on the inner structure of the net. If the net possesses learning or self-organizing ability, it can modify T by attending to features of the external world.

Whenever the condition

$$y = Tx$$

is satisfied, we say that the net has associated pattern $y$ with pattern $x$. We can say that the net recalls $y$ from $x$ or that the net has an ability of coding $x$ into $y$. When k pairs of stimulus patterns $(x_1, z_1), (x_2, z_2), \ldots, (x_k, z_k)$ are presented, the net is required to self-organize by learning these stimulus pairs seriatim such that its transformation T comes to satisfy

$$TX_\alpha = Z_\alpha$$

for all $\alpha = 1,2,\ldots,k$. This type of association or coding has been studied by many theoreticians (Anderson, 1972; Kohonen, 1972; Nakano, 1972; Wigström, 1973). Our task now is to search for a neural mechanism capable of explaining this kind of association or coding.

Pattern generation and concept formation, on the other hand, are the products of a multistable neural net with recurrent connections. The net is in one of several possible inner states and possesses a state-transition mechanism. The change from the present state X to the next state X' is determined by

$$X' = TX,$$

where T denotes the state transition operator of the net. A state X which satisfies

$$X = TX$$

is an equilibrium or stable state. A net may have many stable states.

A pattern generator operates as follows: Given k patterns $X_1, X_2, \ldots, X_k$ which the net is required to generate, the net self-organizes by learning these k patterns such that

$$TX_\alpha = X_\alpha$$

holds for all $X_\alpha$, $\alpha = 1,2,\ldots,k$. It is further required that when a pattern X that is similar to some pattern $X_\alpha$ arrives at the net, which is in state X, the net tends to state $X_\alpha$ and remains in this state. The net thus generates the exact pattern $X_\alpha$ from an approximate pattern X and retains it for a time.

In the case of concept formation, the net receives many patterns from the external world. These patterns form a number of pattern clusters, each cluster corresponding to a concept. In the course of learning these patterns, the net is required to self-organize so that the core pattern of each cluster becomes a stable state of the net. Therefore, when the input patterns are distributed around k core patterns $x_1,\ldots,x_k$, the net modifies itself such that

$$x_\alpha = Tx_\alpha$$

holds for all $x_\alpha$, $\alpha = 1,2,\ldots,k$. After k concepts have thus been formed in the net, the net enters state $x_\alpha$ whenever it receives a pattern $x$ belonging to the cluster of $x_\alpha$, hence understanding pattern $x$ with the help of concept $x_\alpha$. Keeping a pattern in its equilibrium state corresponds to short-term memory, while the formation of stable states corresponds to long-term memory.

In any of the above self-organizing nets, information about many patterns or pattern pairs is stored in a multiply superposed and distributed manner in the synaptic weights of the net. We must now search for a common neural mechanism that embodies these principles of self-organization.

A transformation net for association or coding is structured as shown in Figure 17. It consists of m elements. Each element

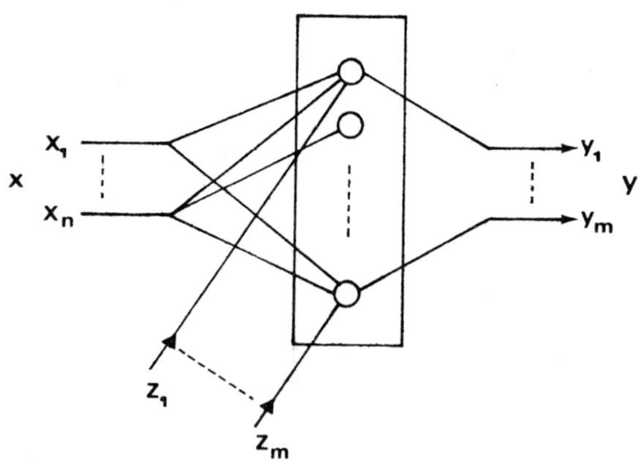


*Figure 17* A transformation net for association.

is a bistable neuron pool having a high and a low activity level. However, we can simplify and approximate the behavior of the neuron pool by using a simple McCulloch-Pitts neuron whose output assumes only two values, namely, 0 and 1. The net has n input lines $x_1, \ldots, x_n$ through which input pattern $\mathbf{x}$ arrives. An input line $x_i$ is connected to all the elements with modifiable synaptic weights. Let $w_{ji}$ denote the synaptic weight of the connection from the *i*th input to the *j*th element. The output from the *j*th element is denoted by $y_j$ with $y_1, \ldots, y_m$ constituting the output pattern

$$\mathbf{y} = \begin{bmatrix} y_1 \\ \vdots \\ y_m \end{bmatrix}.$$

Let $h_i$ represent the threshold of the *i*th element such that

$$\mathbf{h} = \begin{bmatrix} h_1 \\ \vdots \\ h_m \end{bmatrix}.$$

The input/output relation of the net is described as

$$y_i = f\left(\sum_{j=1}^{n} w_{ij} x_j - h_i\right), \qquad i = 1, 2, \ldots, n,$$

or, in vector-matrix notation, as

$$\mathbf{y} = f(\mathbf{W}\mathbf{x} - \mathbf{h}), \tag{4.1}$$

where $\mathbf{W} = (w_{ij})$ is the matrix of the synaptic weights and the step function f operates componentwise. This is the explicit form of the transformation T.

The net has another set of input lines $z_1, \ldots, z_n$. The *i*th element receives input $z_i$ directly. These inputs function as a "teacher" for association or coding. When the net learns a pattern pair $(\mathbf{x}_\alpha, \mathbf{z}_\alpha)$, $\mathbf{x}_\alpha$ enters as characteristic input while $\mathbf{z}_\alpha$

arrives as additional input. In this situation, the synaptic weights are modified.

A net for pattern generation or concept formation can be obtained from a transformation net feeding back the output to the input (Figure 18). Let $w_{ij}$ be the synaptic weight with which the

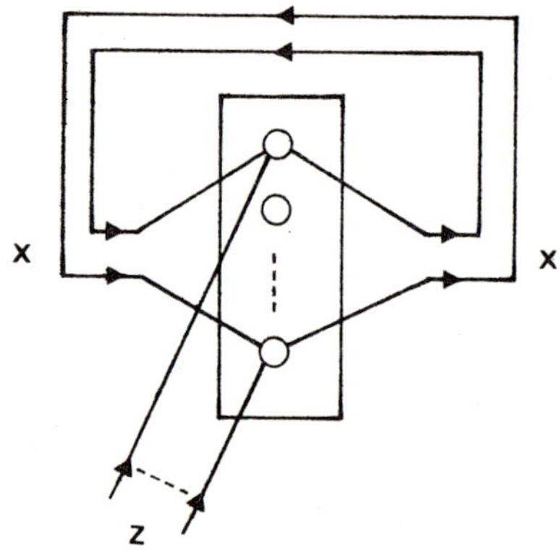


*Figure 18* A net for concept formation.

output of the *j*th element is connected to the *i*th element. The matrix of synaptic weights $\mathbf{W} = (w_{ij})$ is an n x n square matrix because the number of input lines n is equal to the number of output lines. When the net is in state $\mathbf{x}$, the output signal is also $\mathbf{x}$, which determines the next state of the net. The next state $\mathbf{x}' = T\mathbf{x}$ of $\mathbf{x}$ is determined by

$$T\mathbf{x} = f(\mathbf{W}\mathbf{x} - \mathbf{h}). \tag{4.2}$$

In this case the function of the second input $\mathbf{z}$ is to initialize the state of the net. When $\mathbf{z}$ arrives, the net is forced to state $\mathbf{z}$. The net then begins to obey its state-transition dynamics. A distribution of pattern signals $\mathbf{x}$'s from which the net learns also arrive via the second input terminal, forcing the net to assume the respective states. No extra "teachers" exist in this case.

*4.2 Orthogonal and Covariance Learning*

It is generally assumed that a nerve net self-organizes by modifying its synaptic weights $w_i$ (Figure 19). In the Hebbian model of self-organization, $w_i$ increases by a small amount whenever an impulse arrives at that synapse and the post-synaptic neuron fires. To avoid having a synaptic weight that increases without limit, we usually assume that the synaptic weight decays very slowly during the modification process.

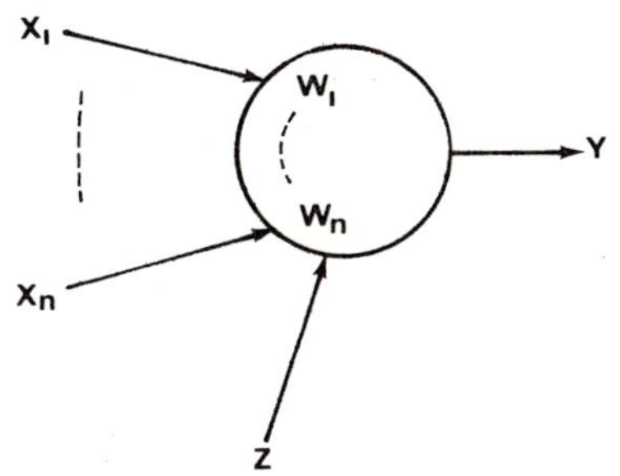


*Figure 19* A neuron with modifiable synapses.

Let us consider a generalized learning rule: A synaptic weight increases in proportion to a reinforcement signal r and decays with a fixed rate. We then have

$$w_i(t + 1) = (1 - \alpha)\, w_i(t) + \beta r_i(t), \tag{4.3}$$

where

$\alpha$ = the decay constant,

$\beta$ = a small learning constant,

$w_i(t)$ = the synaptic weight at time t, and

$r(t)$ = the reinforcement signal to that synapse at time t.

The reinforcement signal generally depends on the input **X**, the output y, the synaptic weights **W**, and the "teacher" signal z if it exists. The Hebbian rule is a special case of (4.3) in which

$r_i$ is given by

$$r_i(t) = x_i(t)y(t).$$

The above signal is equal to 1 whenever a pulse arrives at that synapse and the neuron fires and is 0 otherwise.

We now consider the case where there exists a function $R(W,X,z)$ which yields

$$\frac{\partial R}{\partial w_i} = w_i - \frac{\beta}{\alpha} r_i(W,X,z).* \tag{4.4}$$

In this case, the learning rule of the synaptic weight $w_i$ is expressed as

$$w_i(t+1) = w_i(t) - \alpha \frac{\partial R[W(t),X(t),z(t)]}{\partial w_i}$$

or, in vector notation, as

$$W(t+1) = W(t) - \alpha \frac{\partial R}{\partial W}, \tag{4.5}$$

where $\partial R/\partial W$ denotes the gradient of R with respect to W.

We assume that the input sequence $[X(t),z(t)]$, $t = 1,2,3,\ldots,$ under which the neuron learns is a stationary--or, more precisely, ergodic--random sequence. Let

$$L(W) = \langle R(W,X,z)\rangle$$

be the average of $R(W,X,z)$ over X and z. Moreover, let us assume that the function $L(W)$ has a unique minimum $W_0$. Then the theory of stochastic approximation (see, e.g., Wasan, 1969) guarantees that $W(t)$ of (4.5) converges to the minimum point of $W_0$ of L to within a small fluctuating term which vanishes as $\alpha$ tends to 0.

---

*The necessary and sufficient condition for the existence of R is

$$\frac{\partial r_i}{\partial w_j} = \frac{\partial r_j}{\partial w_i}$$

which holds for most interesting cases.

Therefore, when $\alpha$ is sufficiently small, the synaptic weights of the neuron converge to the minimum of $L(W)$, satisfying the equation

$$\frac{\partial L(W)}{\partial W} = 0$$

or, more explicitly, to that satisfying

$$W = \frac{\beta}{\alpha} \langle r(W,X,z) \rangle, \tag{4.6}$$

where $\langle \; \rangle$ denotes the average over inputs $(X,z)$. By assuming an appropriate function R or reinforcement signal r, we can formulate a learning rule for synapses. We propose here that synaptic modification is governed by orthogonal and covariance learning.

We now generalize this theory so as to include a more realistic case where an input signal X has both excitatory and inhibitory pathways to the neuron (Figure 20). Let $W^+$ be the synaptic

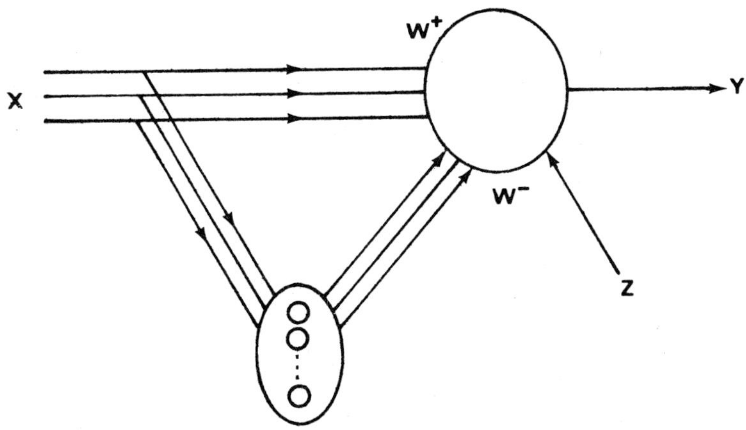


*Figure 20* A neuron having excitatory and inhibitory pathways.

weights of the excitatory pathway and $W^-$ be the synaptic weights of the inhibitory pathway through an inhibitory interneuron pool. The overall connection weights of input X then become

$$W = W^{+} - W^{-}. \tag{4.7}$$

There is no reason to assume that excitatory and inhibitory synapses are subject to the same learning rule. We assume that the reinforcement signal is $x_i z$ for an excitatory synapse (i.e., the synaptic weight increases whenever an impulse arrives simultaneously at the synapse and at the "teacher" signal z) and that the reinforcement signal is

$$x_i \left( \sum_{j=1}^{n} w_j x_j \right).$$

for an inhibitory synapse (i.e., the synaptic weight increases in proportion to the membrane potential $u = \Sigma\, w_j x_j$ only when an impulse arrives at that synapse).

The learning equations are written as

$$W^{+}(t + 1) = (1 - \alpha)\ W^{+}(t) + \beta z X,$$

$$W^{-}(t + 1) = (1 - \alpha)\ W^{-}(t) + \beta'(W \cdot X)X.$$

When

$$\alpha = \alpha'$$

and (4.8)

$$\beta = \beta',$$

by subtracting the second equation from the first we find that

$$W(t + 1) = (1 - \alpha)\ W(t) + \beta(z - W \cdot X)X.$$

The W(t) converges in this case to

$$W_0 = \varepsilon \langle (z - W_0 \cdot X)X \rangle$$

or, by a simple calculation, to

$$W_0 = \langle zX^{T} \rangle (\varepsilon E_n + \langle XX^{T} \rangle)^{-1}, \tag{4.9}$$

where $E_n$ is the n-dimensional unit matrix,

$$\varepsilon = \frac{\beta}{\alpha},$$

and $X^T$ is the transposition of the column vector X; hence, $XX^T$ is the n x n matrix whose (i,j) entry is $x_i x_j$. We can prove that, even when (4.8) is not satisfied, (4.9) holds except for a constant factor (see Amari, 1977b).

Let us now consider the case in which k input signals $X_1, X_2, \ldots, X_k$ are presented to the neurons, and the "teacher" signal $z_\alpha$ is associated with $X_\alpha = 1, 2, \ldots, k$. We also assume, for simplicity's sake, that k vectors arrive at the neuron with equal probability and that they are linearly independent. We then have

$$\langle zX^T \rangle = \frac{1}{k} \sum_{\alpha=1}^{k} z_\alpha X_\alpha^T$$

and

$$\langle XX^T \rangle = \frac{1}{k} \sum_{\alpha=1}^{k} X_\alpha X_\alpha^T.$$

In this case

$$L(W) = \frac{1}{2k} \sum_{\alpha=1}^{k} [|W|^2 + \varepsilon(z_\alpha - W \cdot X_\alpha)^2].$$

Let $X_1^*, \ldots, X_k^*$ be the dual orthogonal vectors of $X_1, X_2, \ldots, X_k$. The dual orthogonal vectors are row vectors defined as follows:

i. $X_\alpha^*$ is a linear combination of $X_1^T, \ldots, X_k^T$.

ii. $X_\alpha^*$ satisfies

$$X_\alpha^* X_\beta = \begin{cases} 1 & \alpha = \beta, \\ 0 & \alpha \neq \beta. \end{cases} \tag{4.10}$$

Using these vectors, we can prove that (4.9) becomes

$$\mathbf{w}_0 = \sum_{\alpha=1}^{k} z_\alpha \mathbf{x}^*_\alpha - \frac{k}{n} \varepsilon \sum_{\alpha\beta} z_\alpha G^{-1}_{\alpha\beta} \mathbf{x}^*_\beta + O(\varepsilon^2), \tag{4.11}$$

where $G^{-1}_{\alpha\beta}$ is the $(\alpha,\beta)$-element of the inverse of the $k \times k$ matrix

$$G = (\mathbf{x}_\alpha \cdot \mathbf{x}^T_\beta), \tag{4.12}$$

and $O(\varepsilon^2)$ is the term of order $\varepsilon^2$.

When $\varepsilon$ is small, (4.11) approximates

$$\mathbf{w}_0 = \sum z_\alpha \mathbf{x}^*_\alpha . \tag{4.13}$$

After the learning is completed, an input pattern $\mathbf{x}_\beta$ arriving at the neuron yields the potential of the amount

$$\mathbf{w}_0 \cdot \mathbf{x}_\beta = z_\beta$$

because of the orthogonality of $\mathbf{x}_\alpha$ and $\mathbf{x}^*_\beta$. This is the signal associated with that input. This learning rule has the ability of orthogonalizing a given set of input patterns $\mathbf{x}_1,\ldots,\mathbf{x}_k$ so that the superposed information can easily be separated. We call the learning rule orthogonal learning. Orthogonal learning is a neural implementation of Kohonen's generalized inverse-matrix method for association (Kohonen, 1974; Kohonen and Oja, 1976).

We also propose another learning rule called covariance learning. The input $\mathbf{x}$ also has an inhibitory pathway. It is therefore possible for the overall synaptic weights to be modified by

$$\mathbf{w}(t+1) = (1-\alpha)\,\mathbf{w}(t) + \beta(z-b)\,(\mathbf{x}-\mathbf{a}), \tag{4.14}$$

where $\mathbf{a}$ is a constant vector and $b$ is a constant. The synaptic weight vector converges to

$$\mathbf{w}_0 = \varepsilon \langle (z-b)(\mathbf{x}-\mathbf{a}) \rangle,$$

which implies that

$$\mathbf{w}_0 = \varepsilon \sum_{\alpha=1}^{k} (z_\alpha - b)\,(\mathbf{x}_\alpha - \mathbf{a}). \tag{4.15}$$

When $a = 0$ and $b = 0$, the rule reduces to the so-called correlation learning principle (see, e.g., Anderson, 1972; Nakano, 1972; Wigström, 1973). When $a$ and $b$ are equal to the average of $x$ and $z$, respectively, $W$ represents the covariances between two signals $x$ and $z$.

### *4.3 Performance of Learning Nets*

When a transformation net (Figure 17) receives repeatedly and equifrequently k pairs of patterns $(x_\alpha, z_\alpha)$, and when the synaptic weights $w_{ij}$'s of all the elements are subject to orthogonal learning, the synaptic weights eventually converge to

$$W = z_\alpha x_\alpha^* - \frac{k}{n} \varepsilon \sum_{\alpha\beta} z_\alpha G_{\alpha\beta}^{-1} x_\beta^* + O(\varepsilon^2). \tag{4.16}$$

After learning is completed, the potential for an input pattern $x_\beta$ is given as

$$u = Wx_\beta = z_\beta - \frac{k}{n} \varepsilon n, \tag{4.17}$$

where

$$n = \sum_{\alpha=1}^{k} z_\alpha G_{\alpha\beta}^{-1} . \tag{4.18}$$

When $\varepsilon = 0$, $u$ coincides with the pattern $z_\beta$ which the net is expected to output. Hence, whenever $\varepsilon$ is small enough to satisfy

$$\left| \frac{k}{n} \varepsilon n_i \right| < 0.5,$$

where $h_i = 0.5$, no component of $(u - h)$ ever changes its sign, even if the term $n$ is added. Hence, the net produces exactly the desired output

$$f(Wx_\beta - h) = z_\beta, \qquad \beta = 1,\ldots,k.$$

Let us consider a more general situation to evaluate the noise immunity of association. We say that a pattern $\tilde{x}_\beta$ is a noisy version of $x_\beta$ with noise level $\delta$ when the rate at which the components of the patterns differ is $\delta$; i.e.,

$$\delta = \frac{1}{n} \sum_{i=1}^{n} |x_{\beta i} - \tilde{x}_{\beta i}|, \tag{4.19}$$

where $x_{\beta i}$ and $\tilde{x}_{\beta i}$ are the *i*th components of $\mathbf{x}_\beta$ and $\tilde{\mathbf{x}}_\beta$, respectively. When $\mathbf{x}_\beta$ with noise level $\delta$ is input to the net, the net outputs

$$\tilde{\mathbf{z}}_\beta = f(W\tilde{\mathbf{x}}_\beta - \mathbf{h}),$$

a noisy version of $\mathbf{z}_\beta$. Let $\delta'$ be the noise level of the output $\tilde{\mathbf{z}}_\beta$. When $\delta'$ is much smaller than $\delta$, association has the effect of reducing the noise. If there is a relation

$$\delta' = \phi(\delta)$$

between $\delta$ and $\delta'$, this demonstrates the noise immunity of an association.

The noise immunity relation depends on various parameters, one of which is

$$s = \frac{k}{n},$$

the ratio of the number of superposed pattern pairs to n. Let a represent the average firing rate of $x_i$ and b the average firing rate of $z_i$. Assuming that $\delta$ and $\varepsilon$ are small, we can then derive the noise immunity relation of association by orthogonal learning; viz.,

$$\delta' \le F\left( \frac{a(1-a)\ (1 - \varepsilon s - \delta)}{2\sqrt{\text{abs}\delta\ (1 - a + s + sa)}} \right), \tag{4.20}$$

where

$$F(u) = \int_u^\infty \frac{1}{\sqrt{2\pi}} \exp\left\{ -\frac{x^2}{2} \right\} dx. \tag{4.21}$$

The result is represented in Figure 21.

When $\varepsilon$ is small, orthogonal learning works well. However, the training pattern pairs $(\mathbf{x}_\alpha, \mathbf{z}_\alpha)$ usually include random noise as well. It has been shown (Amari, 1977b) that this noise has the

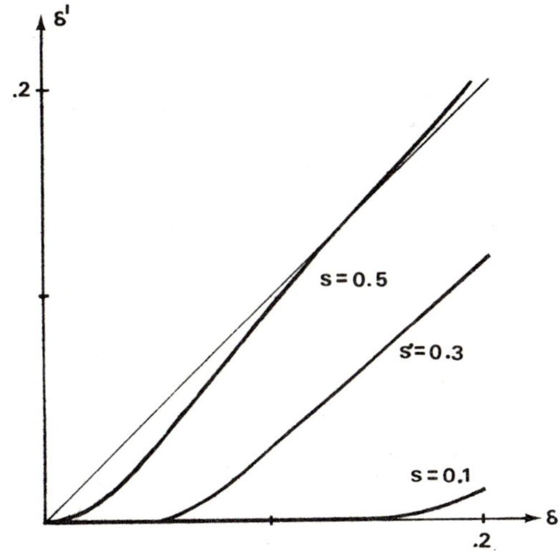


*Figure 21* Noise immunity of association with orthogonal learning.

effect of apparently increasing $\varepsilon$. Hence, when the training pattern pairs include a great deal of noise, the net does not function very well.

In the case of the covariance learning,

$$\delta' = F\left(\frac{(1 - 2\delta)\sqrt{a(1 - a)}}{2\sqrt{sb(1 - b)\ (a + \delta - 2a\delta)}}\right) \tag{4.22}$$

which holds even for large values of $\delta$. However, this evaluation is derived under the assumption that the pattern sets $\mathbf{X}_\alpha$'s and $\mathbf{Z}_\alpha$'s are randomly generated. The results of (4.22) are shown in Figure 22. Covariance learning is inferior to orthogonal learning, but it holds even when the training patterns are very noisy.

We can perform a similar analysis for pattern generation and concept formation in which either orthogonal or covariance learning can be incorporated. The net, in this case, has no special "teacher" signal **Z**. The input signal **X** itself functions as the "teacher" signal. We can prove that the synaptic weights **W**

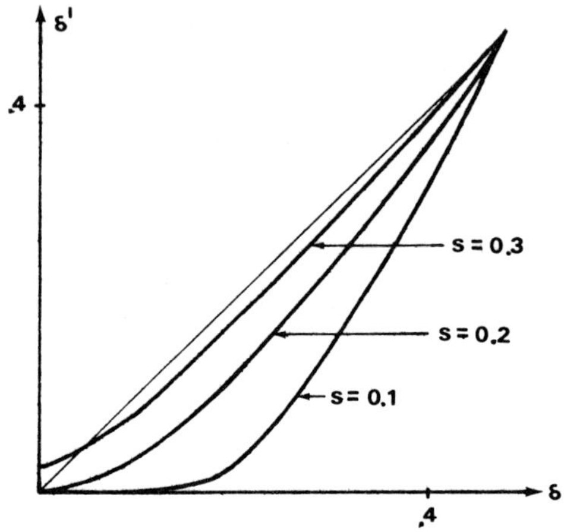


*Figure 22* Noise immunity of association with covariance learning.

converges to

$$W = \sum_{\alpha=1}^{k} \mathbf{x}_\alpha \mathbf{x}^*_\alpha + O(\varepsilon) \tag{4.23}$$

under orthogonal learning. Consequently, when $\varepsilon$ is small,

$$f(Wx_\beta - h) = x_\beta$$

holds for all $\beta = 1,2,\ldots,k$. We observe that $x_\beta$'s have become the stable equilibrium states of the net.

Orthogonal learning, however, does not work well when the input patterns $x_\alpha$ are very noisy. In the case of concept formation, the input patterns may be widely distributed around k core concept patterns and, hence, may produce a great deal of noise.

In the case of covariance learning, the noise immunity relation of the net is given by

$$\delta' \le \phi(\delta) = F\left(\frac{(0.5 - \delta)\ (0.5 - \delta_\alpha)^2}{\sqrt{s(0.5 + \delta)}}\right), \tag{4.22}$$

where $\delta_\alpha$ is a quantity representing the dispersion of the cluster around $\mathbf{x}_\alpha$. This relation holds for a pattern around $\mathbf{x}_\alpha$. When the net is in a state with noise level $\delta$, it subsequently enters a state with noise level $\delta'$. Therefore, the noise level $\delta_t$ at time t satisfies

$$\delta_{t+1} \leq \phi(\delta_t).$$

Since $\phi$ is a monotonically increasing function, we can show that when the equation

$$\delta_{t+1} = \phi(\delta_t)$$

has a stable equilibrium at $\delta = 0$, $X_\alpha$ is a stable equilibrium of the net since the noise level decreases until it converges to 0. Actually, the equation does not have a stable equilibrium at $\delta = 0$, but $\delta = \tilde{\delta}$ where $\tilde{\delta}$ is small. Consequently, the state of the net converges to a state whose noise level is smaller than $\tilde{\delta}$. Thus, we may say that $X_\alpha$ becomes an equilibrium of the net with noise level $\tilde{\delta}$. The behavior of the net is shown in Figure 23. We can see that the concept-formation net works well when s is not very large.

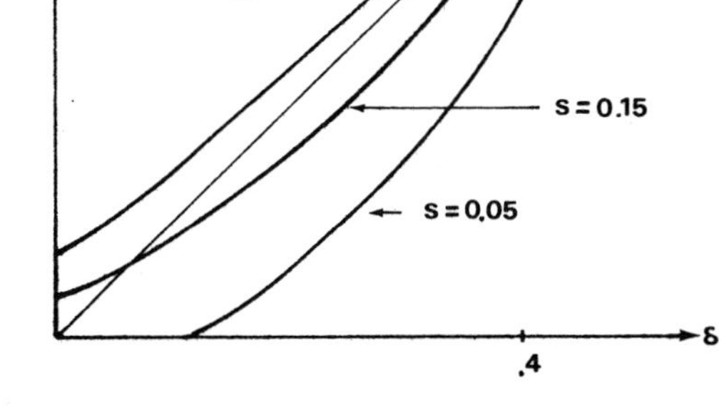


*Figure 23* Noise immunity of concept-formation with covariance learning.

## 5. *CONCLUSION*

We have presented a mathematical theory of neural systems at three different levels; namely, the macroscopic dynamics of neuron pools, the dynamics of pattern formation in neural fields, and the information-processing capabilities of learning nerve nets. In order to make the mathematical analysis tractable and to obtain clear insight into the mechanisms underlying the behaviors of neural systems, we have simplified the behavior of the model neurons or neuron pools as much as possible. Most of our results and much of our methodology remain valid for more realistic models at each of our three levels of analysis. Future research will apply these techniques to the mathematical study of more realistic models of the brain.

ACKNOWLEDGMENTS *It is my pleasure to thank the members of the Center for Systems Neuroscience at the University of Massachusetts at Amherst for their kind help during my tenure as a CSN Fellow. I would also like to express my sincere gratitude to Dr. Jacqueline Metzler, who not only encouraged me to write this article but also provided many valuable suggestions and patiently corrected my English. Special thanks also go to Mrs. Joyce Rodriquez for typing an earlier draft of this manuscript. This research was supported in part by a grant from the Alfred P. Sloan Foundation to the Center for Systems Neuroscience.*

## *REFERENCES*

Ahn, S. M., and Freeman, W. J. 1974. Steady state and limit cycle activity of neurons forming simple feedback loops. I. Lumped circuit model. *Kybernetik 16:* 87-91.

Allanson, J. T. 1956. Some properties of randomly connected neural networks. In C. Cherry, ed., *Proceedings of the Third London Symposium on Information Theory*. London: Butterworths.

Amari, S. 1971. Characteristics of randomly connected threshold element networks and network systems. *Proceedings of the IEEE 59:* 35-47.

Amari S. 1972a. Characteristics of random nets of analog neuron-like elements. *IEEE Transactions on Systems, Man and Cybernetics SMC-2:* 643-657.

Amari, S. 1972b. Learning patterns and pattern sequences by self-organizing nets of threshold elements. *IEEE Transactions on Computers C-21:* 1197-1206.

Amari, S. 1974a. A method of statistical neurodynamics. *Kybernetik 14:* 201-215.

Amari, S. 1974b. A mathematical theory of nerve nets. In M. Kotani, ed., *Advances in Biophysics*. Toyko: University of Tokyo Press.

Amari, S. 1975. Homogeneous nets of neuron-like elements. *Biological Cybernetics 17:* 211-220.

Amari, S. 1977a. Dynamics of pattern formation in a lateral-inhibition type neural field (in preparation).

Amari, S. 1977b. Neural theory of association and concept formation. *Biological Cybernetics* (in press).

Amari, S., Yoshida, K., and Kanatani, K. 1976. A mathematical foundation for statistical neurodynamics. *SIAM Journal on Applied Mathematics* (in press).

Anderson, J. A. 1972. A simple neural network generating interactive memory. *Mathematical Biosciences 14:* 197-220.

Annios, P. A., Beek, B., Csermely, T. J., and Harth, E. M. 1970. Dynamics of neural structures. *Journal of Theoretical Biology 26:* 121-148.

Ashby, W. R., Foerster, F. V., and Walker, C. C. 1962. Instability of pulse activity in a net with threshold. *Nature 196:* 561-562.

Bard, J., and Lauder, I. 1974. How well does Turing's theory of morphogenesis work? *Journal of Theoretical Biology 45:* 501-531.

Ellias, S. A., and Grossberg, S. 1975. Pattern formation, contrast control, and oscillations in the short-term memory of shunting on-center, off-surrounding networks. *Biological Cybernetics 20:* 69-78.

Hadler, K. P. 1974. On the theory of lateral inhibition. *Kybernetik 14:* 161-165.

Kohonen, T. 1972. Correlation matrix memories. *IEEE Transactions on Computers C-21:* 353-359.

Kohonen, T. 1974. An adaptive associative memory principle. *IEEE Transactions on Computers C-23:* 444-445.

Kohonen, T., and Oja, E. 1976. Fast adaptive formation of orthogonalizing filters and associative memory in recurrent networks of neuron-like elements. *Biological Cybernetics 21:* 85-95.

Levin, S. A., ed. 1974. *Some Mathematical Questions in Biology.* Vol. VI. Rhode Island: American Mathematical Society.

Maginu, K. 1975. Reaction-diffusion equation describing morphogenesis. *Mathematical Biosciences 27:* 17-98.

McCulloch, W. S., and Pitts, W. H. 1943. A logical calculus of the ideas imminent in neural nets. *Bulletin of Mathematical Biophysics 5:* 115-133.

Nakano, K. 1972. Association--a model of associative memory. *IEEE Transactions on Systems, Man and Cybernetics SMC-2:* 380-388.

Oguztöreli, M. N. 1975. On the activities in a continuous neural network. *Biological Cybernetics 18:* 41-48.

Reichardt, W., and MacGinite, G. 1962. Zur Theorie der lateralen Inhibition. *Kybernetik 1:* 155-165.

Rozonoer, L. I. 1969a. Random logical nets, I. *Automatika i Telemekhanika 5:* 137-147.

Rozonoer, L. I. 1969b. Random logical nets, II. *Automatika i Telemekhanika 6:* 99-109.

Rozonoer, L. I. 1969c. Random logical nets, III. *Automatika i Telemekhanika 7:* 127-136.

Smith, D. R., and Davidson, C. H. 1962. Maintained activity in neural nets. *Journal of the Association for Computing Machinery 9:* 268-278.

Thom, R. 1972. *Stabilities Structurelles et Morphogenese.* Reading, Massachusetts: W. A. Benjamin.

Turing, A. 1952. The chemical basis of morphogenesis. *Philosophical Transactions, Royal Society of London B-237:* 32-72.

Uesaka, Y., and Ozeki, K. 1972. *Denshi Tsushin Gakkai Ronbunshi* [*Transactions of the Institute of Electronics and Communication Engineers of Japan*] *55D:* 323-330. (Translated in *Systems, Computers, Controls,* 1972, *3(3):* 19.)

Wassan, M. T. 1969. *Stochastic Approximation.* Cambridge: Cambridge University Press.

Wigström, H. 1973. A neuron model with learning capability and its relation to mechanism of association. *Kybernetik 12:* 204-215.

Wilson, H. R., and Cowan, J. D. 1972. Excitatory and inhibitory interactions in local populations of model neurons. *Biophysical Journal 12:* 1-24.

Wilson, H. R., and Cowan, J. D. 1973. A mathematical theory of the functional dynamics of cortical and thalamic nervous tissue. *Kybernetik 13:* 55-80.

Yoshizawa, S. 1974. Some properties of randomly connected networks of neuron-like elements with refractory. *Kybernetik 16:* 173-182.

# Competition and Cooperation in Neural Nets

SHUN-ICHI AMARI*
*Center for Systems Neuroscience*

MICHAEL A. ARBIB
*Department of Computer and Information Science*
*and*
*Center for Systems Neuroscience*

*University of Massachusetts at Amherst*

ABSTRACT *Cooperative and competitive computations seem to play a greater role in decision making in neural circuitry than do executive controls. A number of neural models have been proposed which include cooperative and competitive computations--e.g., the model of the role of the reticular formation in deciding the overall mode of behavior, the model of how the frog's tectum decides the snapping position, and the model of the use of stereopsis to recognize depth in space. The aim of this paper is to develop a mathematical study of these models (especially the latter two) as a theory of neural fields. Special attention is given to the analysis of equilibrium states of the system and their stability, and to the interaction of different stimuli separated in space and time.*

## 1. *ESTABLISHING THE PROBLEM*

One of the first models of decision making in neural circuitry to explicitly opt for cooperative computation, rather than executive control, was the S-RETIC model of Kilmer, McCulloch, and Blum (1969). In modelling the reticular formation of the brainstem,

---

*1975-76 Fellow, on leave from the Department of Mathematical Engineering and Instrumentation Physics, University of Tokyo, Tokyo, Japan.

they used observations that the reticular formation helped switch the organism's gross state from sleep to wakefulness, and vice versa, to suggest that the function of the system was to determine the overall mode of behavior of the organism--such as sleeping, fighting, fleeing, or feeding. From the anatomical work of the Scheibels (1968), they felt that a reasonable structural simplification of the system was as a stack of "poker chips." In their model, each module in the stack was to receive a sample of the overall system input and, on the basis of that sampling, to assign weights to the different modes--so that a given time $p_i(\lambda)$ was to denote the weight given by module i, based on the system's input, to the hypothesis that the appropriate mode of the organism was $\lambda$. However, the modules were then to be coupled in such a way that each module would readjust its weights on the basis of activity in other modules to which it was connected. Kilmer and McCulloch were able to suggest a connection scheme which would lead to eventual consensus, with a majority of the modules assigning the greatest weight to a single mode. This is all done without any executive control.

S-RETIC may be regarded as the prototype for a whole class of general models of competition and cooperation which include mechanisms for segmentation and region labelling (see Arbib, 1975, for further discussion). Here we content ourselves with two other specific models of competition and cooperation. Our goal will be to place these two models--developed using computer simulation--into a rigorous framework, using a mathematical analysis of neural dynamics.

One of the best-known results of neurophysiology is the observation of Lettvin, Maturana, McCulloch, and Pitts (1959) that the ganglion cells of the retina include a class of cells called "bug detectors" because these cells fire most vigorously when a stimulus wiggling like a fly is applied in a particular region of the visual field. Ingle (1968) showed that in a certain region around the head the presence of a fly-like stimulus will elicit a snap from the frog--that is, the frog will turn so that its midline is pointed at the "fly," and it will zap the fly with its tongue. When confronted with two "flies," either of which were vigorous enough that alone they would have elicited a snapping response, the frog could exhibit one of three reactions: it could snap at one of them, it could not snap at all, or it might snap at the "average fly."

Didday (1970, 1976) modelled the *tectum*--the visual region of the midbrain which receives information from the ganglion cells, including the outputs of the bug-detectors--as a competition network whose task was to normally let only one strong signal through from the layer of bug-detector signals. However, Ingle's experiments suggested that this process could fail in two realistic ways: In one case, the competition between two strong inputs would be such that each would hold the other down and neither would become strong enough to take control of the system; while in another case, two nearby signals which were properly balanced could get through simultaneously, yielding a movement to their "mean." The essential mechanism here was to provide each region of the tectum, corresponding to a region of the visual field, with a "sameness" cell whose job was to come up with a weighted sum of activity elsewhere and use this to diminish activity within its own region. In this way, all but the strongest loci of activity would be suppressed, and these activities would then be able to "come up out of the noise" to take control.

We turn now to the problem of segmentation on depth cues, noting that depth in space is signalled by the difference of the retinal coordinates of the projection of a point in space upon the two eyes that we call the *disparity*. Barlow, Blakemore, and Pettigrew (1967) and Pettigrew, Nikara, and Bishop (1968) found that the cells in visual cortex are tuned for disparity. Turning from this physiological evidence to some psychological data on depth perception, we may note the work of Julesz (1971). He prepared input to the two eyes in which each eye received a totally random pattern, but in which there were correlations between the inputs to the two eyes. Specifically, different regions in the two inputs were identical save for a shift in position, yielding--then--a different disparity on the two retinae. Although such a pattern for a naive subject can initially appear to be nothing but visual noise, eventually disparity matching takes place and he perceives surfaces at different depths. Presumably, what causes the initial perception of noise is that, in addition to the correct correlation of points in the two retinae, there are many spurious correlations, and computation is required to reduce these spurious correlations.

The solution to this proposed by Dev (1975) (see, also, Arbib, Boylls, and Dev, 1974; Nelson, 1975) was to imagine the cells of a given disparity as forming a population arrayed in a spatial map corresponding to the map of visual direction. Connections between cells were then to be arranged so that nearby cells of a given disparity would be mutually excitatory, whereas cells which were nearby in visual direction but different in disparity would have inhibitory interaction. In this way, the activity of the array would become organized into a pattern in which, in each region of visual direction, cells of only one disparity type were highly active. As a result, the visual input would eventually be organized into a number of distinct surfaces.

This scheme can be represented as in Figure 1c, in which we use a one-dimensional representation of spatial directions and then have a second dimension for disparity. We have competition along the disparity axis and cooperation along the space axis. We can immediately note, with Arbib (1975) and Montalvo (1975), that the same diagram, Figure 1b, describes S-RETIC as well, save that the cooperation dimension is now the array of modules, and the competition is between modes rather than disparities. Finally, we can see, in Figure 1a, that the Didday model can be regarded as the limiting case of this general scheme of competition

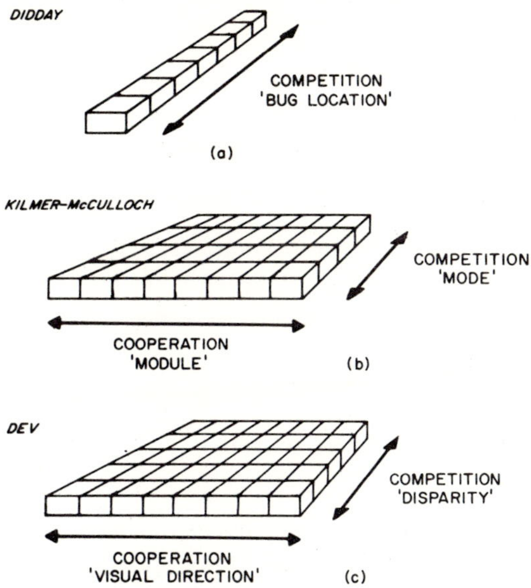


*Figure 1* The Didday model of frog tectum (a), the Kilmer-McCulloch S-RETIC model (b), and the Dev stereopsis model (c) set forth in a common framework of competition and cooperation dimensions. (From Arbib, 1975.*)

and cooperation where there is now only a competition dimension, namely, that of bug location.

Our task in this paper is to provide a mathematical study of networks of the kind shown in Figure 1.

## 2. *DEFINING THE MATHEMATICAL MODEL*

### 2.1 *The Competition Model*

We model the competition dimension by a neuron field connected with an inhibitory neuron pool. In Section 5 we shall show that the field, upon receiving input stimuli, searches for the position of the maximum stimulus at which the frog will snap. For simplicity's sake, we consider a one-dimensional field of excitatory neurons and let $u(x,t)$ be the potential of the field at position x

*M. A. Arbib. 1975. Artificial intelligence and brain theory: Unities and diversities. *Annals of Biomedical Engineering 3:* 258. Reprinted by permission.

and time t. The output of the field at point x is a function $f[u(x,t)]$ of $u(x,t)$ (Figure 2). The field is connected recurrently in such a way that the output of a position gives rise to increases in the potential of other positions. We assume that the field is homogeneous and denote by $w_1(x - \xi)$ the efficiency of this increase in activity at $\xi$ upon the field at x. We may also think of $w_1$ as specifying the strength of connections. Therefore, the strength of stimulation at place x caused by the field $u(x,t)$ is written as

$$\int_{-\infty}^{\infty} w_1(x - \xi) f[u(\xi,t)] d\xi.$$

The structure of the field is assumed to be isotropic; i.e., $w_1(x) = w_1(-x)$. The input stimulation at time t at place x is denoted by $s(x,t)$.

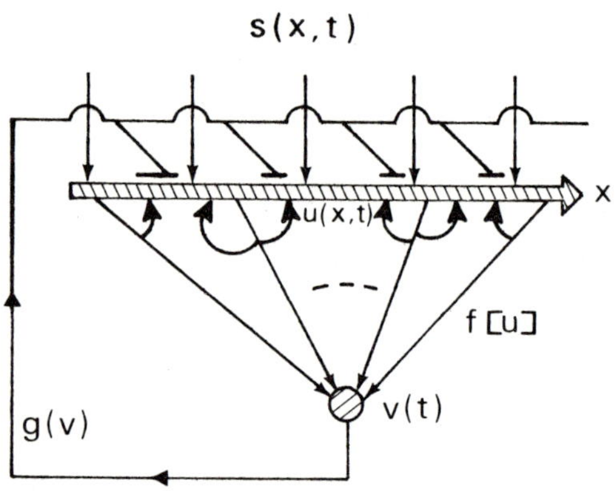


*Figure 2* Competition model.

In this simple model, we do not spatially differentiate the inhibitory neuron pool. The potential and the output activity of this pool are denoted, respectively, by $v(t)$ and $g[v(t)]$. The inhibitory neuron pool receives stimulation from all parts of the excitatory field, and its output, in turn, inhibits all parts of that field with a fixed connection weight $w_2$.

The potential u(x,t) tends to decay to a value $-h_1$ without any stimulation. The time is so normalized that the time constant of this decay equals 1. The potential v(t) of the inhibitory neuron pool decays to a value $-h_2$ with time constant $\tau$. Therefore, the equation of the competition dimension of the present model is written as

$$\left\{\begin{aligned} \frac{\partial u(x,t)}{\partial t} &= -u(x,t) + \int_{-\infty}^{\infty} w_1(x-\xi) f[u(\xi,t)]d\xi \\ &\quad - w_2 g[v(t)] - h_1 + s(x,t), \\ \tau \frac{\partial v(t)}{\partial t} &= -v(t) + \int_{-\infty}^{\infty} f[u(\xi,t)]d\xi - h_2. \end{aligned}\right. \tag{2.1}$$

The output functions f and g are, in general, monotonically increasing, nonlinear functions. The point of the model is the multistability of the excitatory field--a portion of the field, although usually in a low activity level, can retain a high activity level, too, by virtue of the self-excitation through recurrent connections. It has been shown (Amari, 1971, 1972, 1974) that the low and high activity levels of such a multistable (bistable) net are kept nearly constant even if the strength of input stimuli changes (Figure 3). Therefore, by assuming the high level

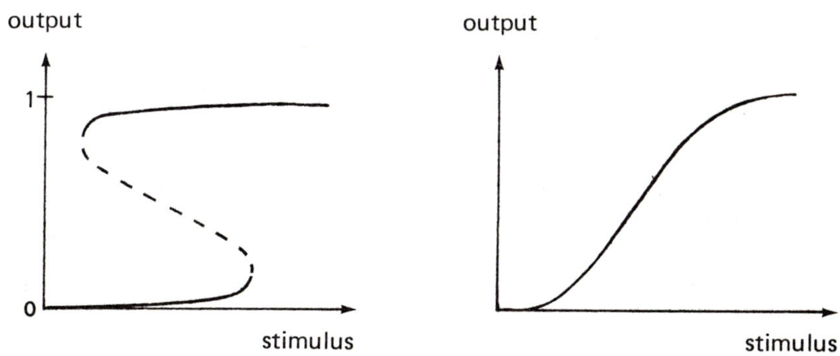


a) bistable b) monostable

*Figure 3* Outputs from a bistable and a monostable neuron pool.

equal to 1 and the low level equal to 0, we may treat the special case where the output function f takes on two values, 0 and 1;

$$f(u) = \begin{cases} 1 & \text{when } u > 0, \\ 0 & \text{when } u \leq 0. \end{cases} \tag{2.2}$$

This mathematical simplification makes analysis of the model easier. However, almost the same results follow even if the function f is a general continuous function, so long as the self-excitation connection is strong enough to guarantee the bistable character. We assume that the inhibitory neuron pool is monostable and has output function

$$g(v) = \begin{cases} v & \text{when } v > 0, \\ 0 & \text{when } v \leq 0. \end{cases} \tag{2.3}$$

### *2.2 The Full Model*

The full model adds a cooperation dimension x to the competition dimension d (corresponding to spatial dimension x and disparity dimension d in the stereopsis model). It consists of a two-dimensional field of excitatory neurons or neuron pools and a one-dimensional field of inhibitory neurons or neuron pools (Figure 4).

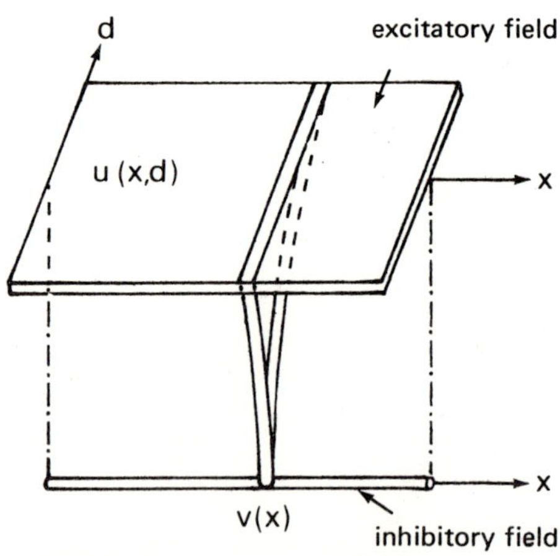


*Figure 4* The full model.

For fixed x, the model is to reduce to the competition model, having only one spatial dimension d. However, in addition to the interaction within an "x-slice," there are *excitatory* interactions between the "slices."

The excitatory field is recurrently connected with a weighting function $w_1(x,d)$ which is the strength of connection from a point to the point distant $(x,d)$. This weighting function of the connections has a broad spread in the cooperation dimension x and a narrow spread in the competition dimension d (Figure 5). The inhibitory field has only the cooperation dimension. It is activated by outputs from the excitatory field of the same spatial positions and inhibits, in turn, the elements of all depths of the spatial positions around its position. Let $w_2(x)$ be the weighting function of the connection from the inhibitory field to the excitatory field. It is also of a very narrow spread.

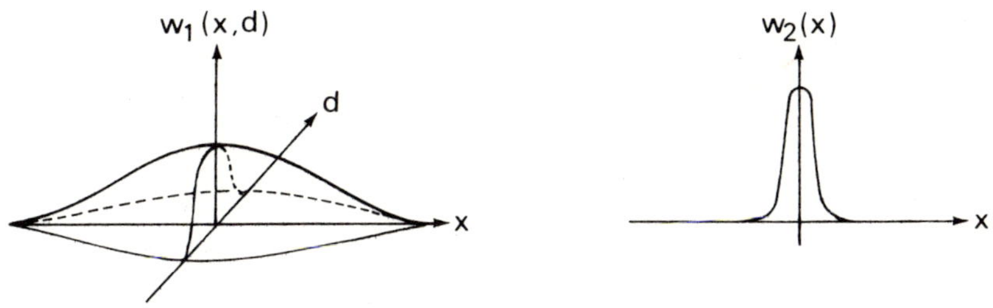


*Figure 5* Weighting functions of connections in the full model.

The model is a natural extension of the preceding competition model, and the equation is written as

$$\left\{\begin{aligned} \frac{\partial u(x,d,t)}{\partial t} &= -\,u(x,d,t) \\ &\quad + \int_{-\infty}^{\infty}\int_{-\infty}^{\infty} w_1(x-\xi,\ d-\eta) f[u(\xi,\eta,t)]\,d\xi d\eta \\ &\quad - \int_{-\infty}^{\infty}\int_{-\infty}^{\infty} w_2(x-\xi) g[v(\xi,t)]\,d\xi - h_1 \\ &\quad + s(x,d,t), \\ \tau\frac{\partial v(x,t)}{\partial t} &= -\,v(x,t) - h_2 + \int_{-\infty}^{\infty} f[u(x,\eta,t)]\,d\eta, \end{aligned}\right. \tag{2.4}$$

where u(x,d,t) and v(x,t) are the potentials of the excitatory field at position x, depth d, time t; and of the inhibitory field at position x, time t, respectively.

Now we analyze a semidiscretized version of the above equation in which the depth dimension is quantized. Then the two-dimensional field splits into many one-dimensional layers. Let $u_d(x,t)$ be the potential at position x at time t of the layer corresponding to the *d*th depth. The equation of this simplified version is written as

$$\begin{cases} \dfrac{\partial u_d(x,t)}{\partial t} = -u_d(x,t) + \displaystyle\int_{-\infty}^{\infty} w_1(x-\xi) f[u_d(\xi,t)]d\xi \\ \qquad - \displaystyle\int_{-\infty}^{\infty} w_2(x-\xi) g[v(\xi,t)]d\xi - h_1 \\ \qquad + s_d(x,t), \qquad d = 1,2,3,\ldots,d_0, \\ \tau \dfrac{\partial v(x,t)}{\partial t} = -v(x,t) - h_2 + \displaystyle\sum_d f[u_d(x,t)], \end{cases} \tag{2.5}$$

where the depth dimension is discretized roughly enough so that no mutually excitatory connections exist between different layers.

We shall refer to (2.1) as the competition model and to (2.5) as the full model. A cooperative process takes place within a layer, while a competitive process takes place among layers in the "full" model. The full model clarifies the cooperation and competition mechanisms well in the case where a given input has piecewise constant depth. When the input has continuously varying depth, we need to treat the original equation (2.4) directly.

The model is anatomically quite similar to that of Dev (1975). The only difference is that the present model is multistable in character while the Dev model is monostable. The multistability is caused by a strong self-excitatory connection and admits the use of a simple two-valued output function f. The multistability makes it possible that, given an input stimulus, the model does not have a unique depth solution but has a number of possible solutions to one of which the model really converges. This gives rise to hysteresis phenomena which are observed in the real visual system.

*2.3 Monolayer Activity in the Full Model*

The full model has no excitatory connections between different layers. Thus, if the net is initially quiescent and input is applied to only one layer, say layer d, then we may restrict our attention purely to the interaction of layer d with the inhibitory layer. We then have

$$\frac{\partial u_d(x,t)}{\partial t} = - u_d(x,t) + \int_{-\infty}^{\infty} w_1(x - \xi) f[u_d(\xi,t)] d\xi - \int_{-\infty}^{\infty} w_2(x - \xi) g[v(\xi,t)] d\xi - h_1 + s_d(x,t),$$

where, near equilibrium, we can approximate $v(\xi,t)$ by $f[u_d(\xi,t)] - h_2$. Now $g[f[u_d(\xi,t)] - h_2]$ vanishes outside the support $R[u_d]$ of $u_d$ and takes the values $1 - h_2$ in $R[u_d]$, since we now assume $1 > h_2$. We thus have, near equilibrium, that a single layer is governed by the equation

$$\frac{\partial u_d(x,t)}{\partial t} = - u_d(x,t) + \int_{R[u_d]} w(x - \xi) d\xi - h_1 + s_d(x,t), \tag{2.6}$$

where $w(x) = w_1(x) - w_2(x)(1 - h_2)$.

We now introduce:

*Assumption 1:* $h_1 > 0$ and $1 > h_2 > 0$.

We then have the following result for the full model, which reduces to a result for the competition model when we ignore d in $u_d(x)$ and x in $v(x)$.

*Theorem 1:* The unexcited state given by

$$u_d(x) = -h_1, \quad v(x) = -h_2$$

is a stable equilibrium for $s_d(x) = 0$. An input stimulation satisfying

$$s_d(x) < h_1$$

for all d and x cannot excite the system from this resting state.

*Proof:* This is immediate from the dynamical equations, since $u_d(x,t)$ never becomes positive in this case. □

The reader will note an apparent paradox: A single layer (*cooperation* dimension) of the full model looks--near equilibrium--like the *competition* model. We shall exploit this similarity in later sections--where we shall see the implications of the differing shapes of $w_1$ and $w$.

Before turning to this analysis of a continuous neuron field, however, we look at a *primitive* competition model--which corresponds to a section across the full model for a fixed x.

### 3. *THE PRIMITIVE COMPETITION MODEL*

We devote this section to the primitive competition model of Figure 6, which consists of n excitatory elements and one inhibitory element. An excitatory element is a neuron pool including self-excitation connections. If we connect neighboring excitatory

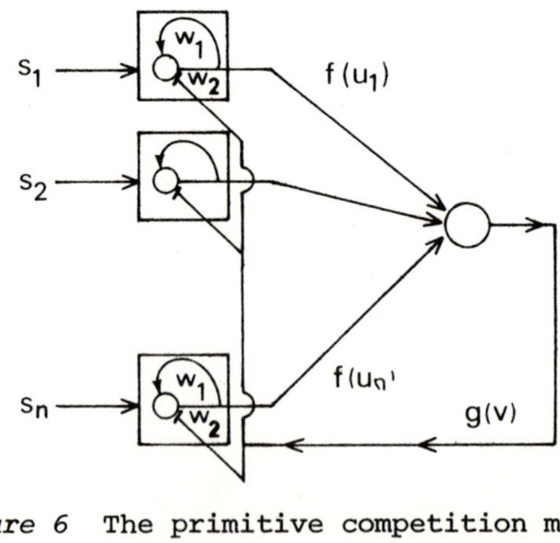


*Figure 6* The primitive competition model.

elements to form a continuous field, we again obtain the competition model.

Let $u_i$ be the average potential of the *i*th element, $s_i$ the input, and v the average potential of the inhibitory element. Then, the behavior of the primitive module can be described by the following set of differential equations:

$$\begin{cases} \dfrac{du_i(t)}{dt} = -u_i + w_1 f(u_i) - w_2 g(v) - h_1 + s_1, \\ \qquad i = 1,\ldots,n, \\ \tau \dfrac{dv(t)}{dt} = -v + \sum\limits_{i=1}^{n} f(u_i) - h_2, \end{cases} \tag{3.1}$$

where

$$f(u) = \begin{cases} 1 & u > 0, \\ 0 & u \le 0, \end{cases}$$

$$g(u) = \begin{cases} u & u > 0, \\ 0 & u \le 0. \end{cases}$$

We will show that, given a set of stimuli $s_1, s_2, \ldots, s_n$, the excitatory elements compete with each other. One element which receives the maximum stimulus eventually wins and remains in the excited state, whereas all the others eventually stay in the quiescent state. The module has a multistable character. If an element, say element 1, has once won, it will remain in the excited state even if another stimulus becomes bigger than $s_1$-- unless it becomes much bigger. Therefore, the module shows hysteresis. We shall show how to add a threshold-controlling routine that can cancel the previous excitation and initialize the module.

### *3.1 Equilibrium States of the Primitive Competition Model*

These satisfy

$$u_i = w_1 f(u_i) - w_2 g[\sum_{j=1}^{n} f(u_j) - h_2] - h_1 + s_1, \qquad i = 1,2,\ldots,n, \tag{3.2}$$

where $s_i$ is kept constant. Now we put the following restrictions on the parameters:

i. $h_1 > 0$,

ii. $0 < h_2 < 1$,

iii. $w = w_1 - (1 - h_2)w_2 > 0$,

iv. $w_2 > w - h_1 + s_{max}$,

where $s_{max}$ is the maximum level of the input. The conditions $h_1 > 0$ and $h_2 > 0$ mean that an element does not spontaneously fire without stimulation. The condition $h_2 < 1$ implies that the inhibitory element is activated by excitation of any one of the excitatory elements. Condition (iii) guarantees the bistable character. Condition (iv) guarantees that only one element can remain excited in an equilibrium. We show these in the following theorem.

*Theorem 2:*

i. The quiescent state in which no elements are excited is a stable equilibrium only when all the stimuli $s_i$ are smaller than $h_1$, $s_i < h_1$ $(i = 1,2,\ldots,n)$; and

ii. At most one element can be excited in the equilibrium.

*Proof:*

i. Assume that no elements are excited in the equilibrium (i.e., each $u_j < 0$). Since all $u_j < 0$, we have $f(u_i) = g[\Sigma f(u_j) - h_2] = 0$, and hence,

$$u_i = - h_1 + s_i.$$

Then $u_i < 0$ implies $s_i < h_1$, as was to be shown.

ii. Assume that at least two elements are excited in the equilibrium. We then have $\sum_j f(u_j) \geq 2$. Since $h_2 < 1$, we then have

$$g[\sum_j f(u_j) - h_2] \geq 2 - h_2.$$

Let i be one of the excited elements ($u_i > 0$). Then

$$\begin{aligned} u_i &= w_1 - w_2 g[\Sigma f(u_j) - h_2] - h_1 + s_i, \\ &\leq w_1 - (2 - h_2)w_2 - h_1 + s_{max}, \\ &= w - w_2 - h_1 + s_{max}, \\ &< 0, \end{aligned}$$

which is a contradiction. □

The theorem shows that $h_1$ is the threshold of activation of the module and that, when some stimuli are bigger than the threshold $h_1$, these elements may be activated. But at most one element can win the competition to remain in the excited state at equilibrium.

### 3.2 *The Dynamics of Competition*

We now show that if an element really "wins," it is the one receiving the maximum stimulus.

*Theorem 3:* If an element remains excited in the equilibrium, it is the one receiving the maximum stimulus, provided all $u_i$'s are initially the same.

*Proof:* Assume that $s_1 > s_2$ and $u_1(0) = u_2(0)$. Then,

$$U(t) = u_1(t) - u_2(t)$$

satisfies

$$\frac{d}{dt} U(t) = - U(t) + w_1[f(u_1) - f(u_2)] + s_1 - s_2.$$

Since the solution U(t) of the equation

$$\frac{dU(t)}{dt} = - U(t) + a(t)$$

never becomes negative if $U(0) \geq 0$ and $a(t) > 0$, and

$$f(u_1) - f(u_2) \geq 0$$

if $u_1 \geq u_2$, we have $U(t) > 0$ or

$$u_1(t) > u_2(t).$$

Since only one element can remain in the excited state in equilibrium, it must be the element which receives the maximum stimulus.□

The theorem does not exclude the possibility that more than two elements are each activated prior to equilibrium. This really occurs, as is shown in Figure 7. When $s_1$ and $s_2$ are bigger than $h_1$ and $s_1 > s_2$, element 1 first begins to be excited. However, by

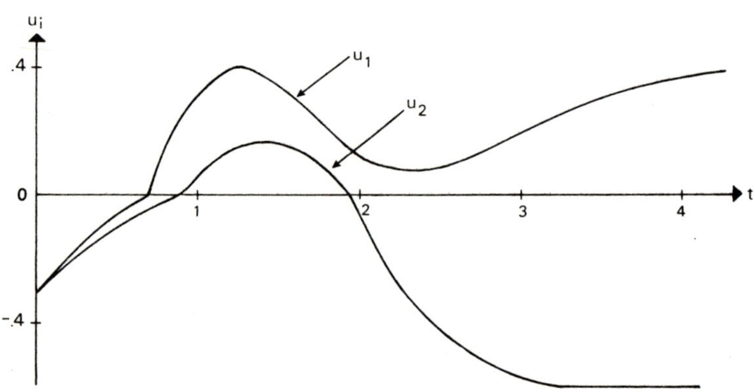


*Figure 7* Competing processes in the primitive model: $w_1 = 1$, $w_2 = 1$, $h_1 = 0.3$, $h_2 = 0.2$, $\tau = 0.5$, $s_1 = 0.6$, $s_2 = 0.5$.

virtue of the time lag of the activation of inhibition, element 2 may become excited, too. Then--even though $u_1(t) - u_2(t)$ is still increasing--both begin to decrease until element 2 becomes inactivated. There are many cases depending on the magnitude of $s_1$ and $s_2$. When $s_1$ is sufficiently large compared to $s_2$, only element 1 is ever excited. When $s_1$ and $s_2$ are sufficiently close, not only do both become excited, but both of them may again become inactivated, after which element 1 again becomes excited. On area A in Figure 8, both elements become activated prior to equilibrium. Area A becomes smaller as $\tau$ becomes smaller. It is also possible for more than two elements to be activated prior to equilibrium in this way.

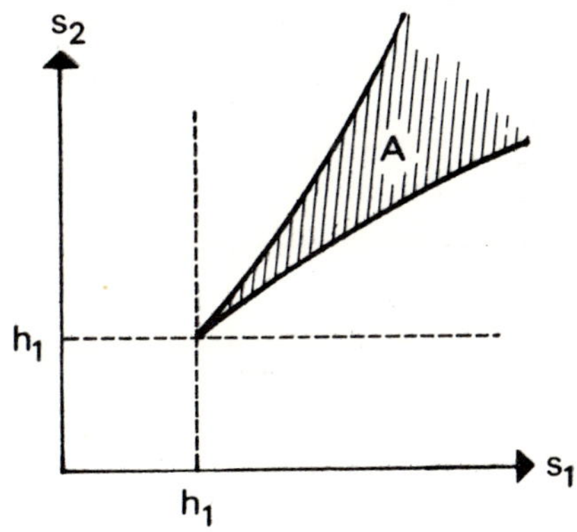


*Figure 8* Region of stimuli, A, which activates two elements prior to equilibrium.

This phenomenon has an interesting meaning in connection with the full model of Section 2, which we may regard as a network in which a number of primitive competition models are interconnected. When a component module receives an array of stimuli, and if one of them is sufficiently strong, only the element of the maximum stimulus becomes activated as the candidate. However, if there are a number of stimuli of similar magnitude, each of their elements becomes activated. The cooperative signal will arrive at this time from the activated elements of the other modules in each of which one or some candidate elements are activated, and the element which receives the maximum sum of the original and cooperative stimuli will eventually win.

### *3.3 Resetting the Network*

The primitive module is multistable and hence exhibits hysteresis. An element enters the excited state when it receives a stimulus bigger than $h_1$. Once it has been excited, it becomes inactivated only when the stimulus becomes smaller than $h_1 - w$ because it receives self-stimulation of amount $w = w_1 - (1 - h_2)w_2$. When one element is excited, a stimulus bigger than $h_1 + (1 - h_2)w_2$ is necessary for another element to be excited because it receives an inhibitory stimulus of strength $(1 - h_2)w_2$. It becomes inactivated when the stimulus becomes smaller than $h_1 - w_1$ (see Figure 9).

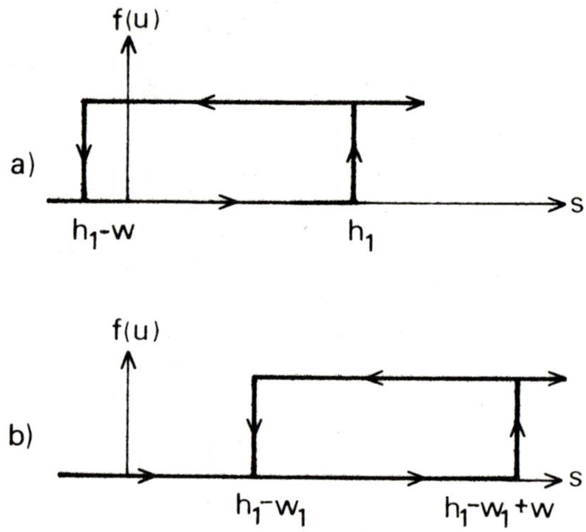


*Figure 9* Hysteresis curves for (a) excited element and (b) non-excited element.

Since the module is multistable, we need to consider the mechanism of transition from one stable state to a more plausible stable state when the input changes. Let element 1 receive the maximum stimulus and become activated. Suppose that stimulus $s_2$ later becomes bigger than $s_1$. Element 1 remains excited once it becomes excited. It is possible for the module to reconsider and to take the new situation into account by changing the threshold value $h_1$. There are two ways of changing the threshold--increasing it or decreasing it.

When $h_1$ changes to a bigger threshold $h_1'$ $(h_1' > h_1)$, element 1 remains excited when $s_1 > h_1' - w$ and is inactivated when $s_1 < h_1' - w$. Once it is inactivated, element 2, which receives a bigger stimulus $s_2$, becomes newly excited once the threshold $h_1'$ returns to the original $h_1$. Therefore, by this process of changing the threshold, $h_1 \to h_1' \to h_1$, the module can take the new situation into account and the alternation of excitation from element 1 to element 2 takes place, provided $s_1$ is smaller than $h_1' - w$. However, if $s_1$ is bigger than $h_1' - w$, no change occurs even if $s_2$ is far bigger.

The other process is the change $h_1 \to h_1'' \to h_1$ $(h_1'' < h_1)$ of the threshold. In this case, element 2 is activated only when

$$s_2 > h_1'' + (1 - h_2)w_2.$$

When element 2 becomes activated, then by increasing the threshold to again equal $h_1$, element 1 becomes inactivated because only one element can be excited at the threshold $h_1$ and $s_1 < s_2$. However, even when $h_1$ changes to $h_1''$, no change occurs if $s_2$ (or other $s_i$'s) is weaker than $h_1'' + (1 - h_2)w_2$. We may call

$$s_c = h_1'' + (1 - h_2)w_2$$

the *challenging value*. By this process, the elements which receive a stimulus bigger than $s_c$ can challenge the element which has been already excited, and the one which receives the maximum stimulus wins among them.

We can thus extend the concept of stability. A state is called, respectively, *T-stable* and *C-stable* if it is stable under the changes of the threshold $h_1 \to h_1' \to h_1$ and $h_1 \to h_1'' \to h_1$.

## 4. *SINGLE-LAYER EQUILIBRIA*

We now return to the models of Section 2 and consider equilibria in the competition model and equilibria involving a single layer in the full model. Let $u(x,t)$ denote the potential in the competition model or in the layer of interest in the full model.

Consider the case shown in Figure 10 in which activity in the layer is positive in a single interval $(x_1,x_2)$ of length a. We assume that this positive peak is sufficiently isolated from other peaks that we may ignore them in the following analysis. (Our analysis follows the technique of Amari, 1977.)

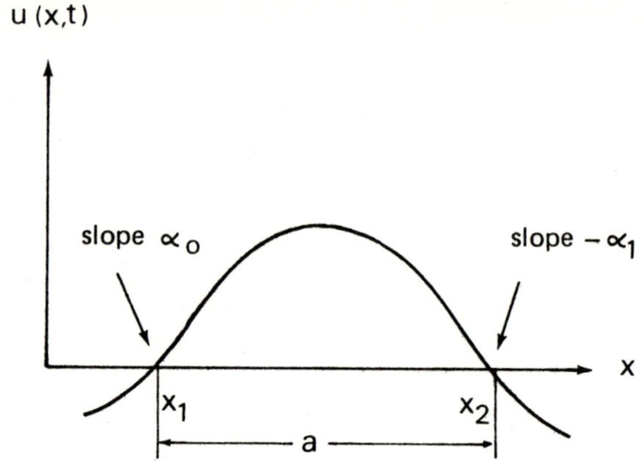


*Figure 10* Single-peaked excitation with length a.

We have that $f[u(x,t)]$ is 1 if x is in $(x_1, x_2)$, and 0 elsewhere, so that

$$\int_{-\infty}^{\infty} w_1(x - \xi) f[u(\xi,t)] d\xi = \int_{x_1}^{x_2} w_1(x - \xi) d\xi$$

$$= W_1(x - x_1) + W_1(x_2 - x)$$

on setting

$$W_1(x) = \int_0^x w_1(\xi) d\xi.$$

We then deduce from (2.1) that

$$\frac{\partial u(x,t)}{\partial t} = - u(x,t) + W_1(x - x_1) + W_1(x_2 - x) - w_2 g[v(t)] - h_1 + s(x,t). \tag{4.1}$$

Similarily, on setting

$$W(x) = \int_0^x w(\xi) d\xi,$$

we obtain from (2.6)

$$\frac{\partial u(x,t)}{\partial t} = - u(x,t) + W(x - x_1) + W(x_2 - x) - h_1 + s(x,t) \tag{4.2}$$

for a single peak of activity in a single layer of the full model near equilibrium.

From the fact that $u(x_i,t) = 0$ and $x_2 - x_1 = a$, we see that, for the full model near equilibrium,

$$\frac{\partial u(x_i,t)}{\partial t} = W(a) - h_1 + s(x_i,t),$$

while, for the competition model, we have

$$\frac{\partial u(x_i,t)}{\partial t} = W_1(a) - w_2 g(v) - h_1 + s(x_i,t).$$

In either case, we have

$$\frac{dx_i}{dt} = - \frac{\partial u(x_i,t)}{\partial t} \Big/ \frac{\partial u(x_i,t)}{\partial x}$$

and

$$\frac{da}{dt} = \frac{dx_2}{dt} - \frac{dx_1}{dt} .$$

Setting

$$c = \frac{1}{\partial u(x_1,t)/\partial x} - \frac{1}{\partial u(x_2,t)/\partial x} = \left( \frac{1}{\alpha_0} + \frac{1}{\alpha_1} \right),$$

we obtain:

i. For the full model near equilibrium,

$$\frac{da}{dt} = c[W(a) - h_1 + s] \tag{4.3}$$

if $s_d(x,t) \equiv s$. Thus at equilibrium,

$$W(a) = h_1 - s. \tag{4.4}$$

ii. For the competition model,

$$\frac{da}{dt} = c[W_1(a) - w_2 g(v) - h_1 + s] \tag{4.5}$$

if $s_d(x,t) \equiv s$. Thus at equilibrium,

$$W_1(a) = w_2 g(v) + h_1 - s. \tag{4.6}$$

We now examine the equilibrium properties of such a region in more detail.

*Theorem 4:* Consider the following conditions on the structure parameters of the competition model. First, we assume that the connection weight function $w_1(x)$ is a symmetric, non-negative, unimodal function with $w_1'(x) \leq 0$ for $x \geq 0$ as is shown in Figure 11, and $w_1(x) = 0$ for $|x| \geq \delta$. The latter means that the maximal length of connections is limited to $\delta$ in the field. We then require:

i. $h_1 > 0$, $h_2 > 0$,

ii. $W_{1\infty} > w_2 g(\delta - h_2) + h_1$,

where $W_{1\infty} = W_1(\infty)$. There are exactly two lengths, $a_0$ and $a_1$, for equilibrium solutions of a connected excitation region in the competition model with zero input. They satisfy

$$a_0 < \delta < a_1.$$

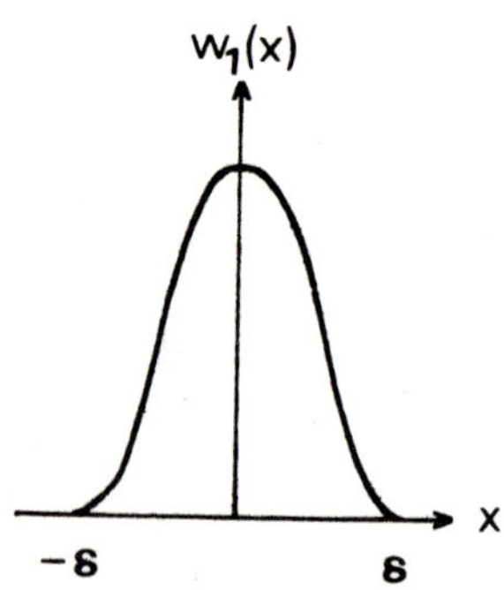


*Figure 11* Weighting function of connections for the competition model.

Moreover, the solution with length $a_1$ is stable and the solution with length $a_0$ is unstable.

*Proof:* The graph of $W_1(a)$ is convex and monotonically increasing starting with the origin, and flat for $a > \delta$, as is shown in Figure 12. On the other hand, the graph of $w_2 g(a - h_2) + h_1$ is flat at the level $h_1$ for $a < h_2$, and increases linearly thereafter. From (ii) we see that, at $a = \delta$,

$$W_1(a) > w_2 g(a - h_2) + h_1.$$

Moreover, at $a = 0$, we have

$$W_1(0) = 0 < h_1.$$

Therefore, the equation (4.6) has exactly two solutions, separated by $\delta$.

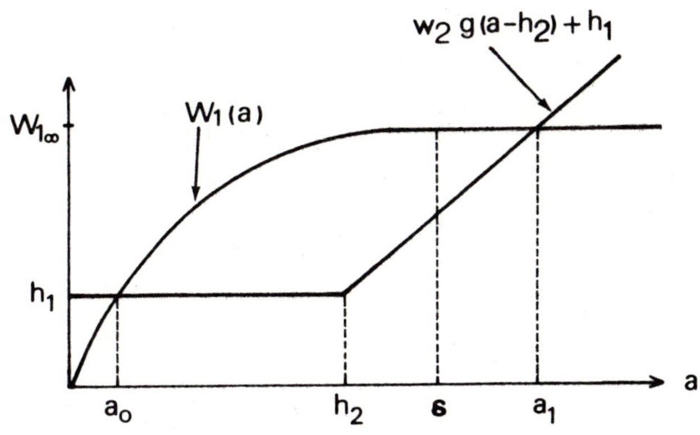


*Figure 12* Graphs of $W_1(a)$ and $w_2 g(a - h_1) + h_1$.

For an excitation pattern of length a and for potential v of the inhibitory neuron pool, the time rate of change in the length of the excited region is given [recall (4.5) with $s \equiv 0$] by

$$\frac{da}{dt} = \left(\frac{1}{\alpha_0} + \frac{1}{\alpha_1}\right) [W_1(a) - w_2 g(v) - h_1],$$

where

$$\alpha_0 = \frac{\partial u(0)}{\partial x},$$

$$\alpha_1 = -\frac{\partial u(a)}{\partial x}.$$

The change of v is described by

$$\frac{dv}{dt} = \frac{1}{\tau}\left(-v + a - h_2\right).$$

In the neighborhood of the equilibrium waveform of length $a_1$,

$$\alpha = \alpha_0 = \alpha_1 = w_1(0)$$

because $w_1'(a_1) = 0$. In order to see the stability of the equilibrium solution, we put

$$\begin{cases} a = a_1 + \varepsilon_1, \\ v = a_1 - h_2 + \varepsilon_2, \end{cases}$$

where $\varepsilon_1$ and $\varepsilon_2$ are small deviations from the equilibrium. We then obtain the linearized variational equation

$$\frac{d}{dt}\begin{pmatrix} \varepsilon_1 \\ \varepsilon_2 \end{pmatrix} = \begin{pmatrix} 0 & -\frac{2}{\infty} w_2 \\ \frac{1}{\tau} & -\frac{1}{\tau} \end{pmatrix}\begin{pmatrix} \varepsilon_1 \\ \varepsilon_2 \end{pmatrix},$$

where

$$\frac{dW_1(a_1)}{da} = w_1(a_1) = 0$$

is taken into account. Since the trace of the coefficient matrix is negative and the determinant is positive, the solution $a_1$ is stable. The instability of the solution $a_0$ is proved similarly. □

The interaction of two excited regions will be treated in Section 6. We now study the equilibrium behavior of a single layer of the full model:

*Theorem 5:* Assume $h_1 > W_\infty = W(\infty) > 0$ and that $1 > h_2 > 0$. Consider the case where a uniform stimulus of intensity s is applied to a layer of the full model, but no stimuli to the other layers. Then,

i. when $s < s_0$, where $s_0 = h_1 - W_\infty$, the layer cannot retain any excitation;

ii. when $s > h_1$, the whole layer becomes excited; and

iii. when $s_0 < s < h_1$, the whole layer is either excited or quiescent, depending on past hysteresis.

*Proof:* Recall that the dynamical equation for the change of the length of the excited region near equilibrium is

$$\frac{da}{dt} = c[W(a) - h_1 + s],$$

where $c > 0$. The full model, hence, has an equilibrium solution of a finite excited region of length a when a satisfies

$$W(a) - h_1 + s = 0.$$

The function W(a) has the shape shown in Figure 13.

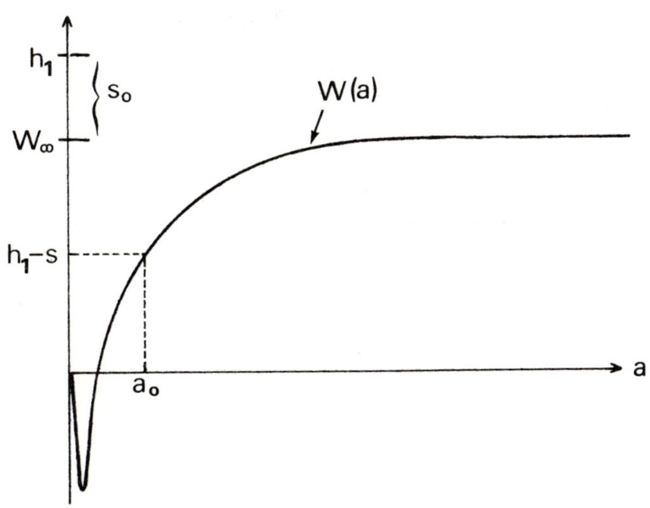


*Figure 13* Graph of W(a).

i. When $s < s_0$, the equation has no solutions and $W(a) - h_1 + s < 0$ or $\frac{da}{dt} < 0$ always holds. This shows that an excited region of any length shrinks and eventually vanishes.

ii. When $s > h_1$, no position of the layer can remain unexcited, so that the whole layer becomes excited.

iii. When $s_0 < s < h_1$, we have the equilibrium solution of length $a_0$. However, this solution is unstable because

$$\frac{da}{dt} > 0 \qquad \text{for } a > a_0,$$

$$\frac{da}{dt} < 0 \qquad \text{for } a < a_1.$$

Hence, if the initial excited region is longer than $a_0$, it grows until the whole layer becomes excited. If the initial region is smaller than $a_0$, it shrinks and vanishes. □

## 5. *RESPONSE TO A PEAK OF INPUT STIMULATION*

We first consider the response of the competition model to the noiseless single-peaked stimulus $s(x)$ shown in Figure 14. We assume that the equilibrium solution $u(x)$ for this input has the

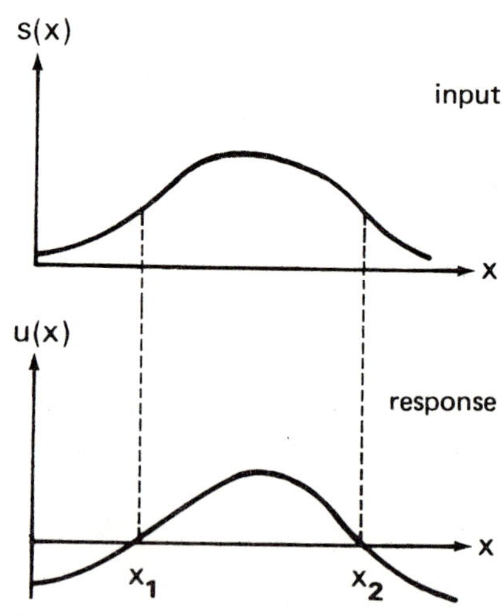


*Figure 14* Response of the competition model to a unimodal stimulus.

unimodal form shown in Figure 14, where $a = x_2 - x_1$, and $R[u] = (x_1, x_2)$. At equilibrium, (4.1) yields

$$u(x) = W_1(x - x_1) + W_1(x_2 - x) - w_2 g(a - h_2) - h_1 + s(x).$$

From

$$u(x_1) = u(x_2) = 0,$$

we have

$$\begin{cases} W_1(a) = w_2 g(a - h_2) + h_1 - s(x_1), \\ W_1(a) = w_2 g(a - h_2) + h_1 - s(x_2), \\ \quad a = x_2 - x_1. \end{cases} \tag{5.1}$$

This implies

$$s(x_1) = s(x_2),$$

i.e., the intensity of stimulation is the same at the boundaries. In the case of a single-peaked, noiseless stimulus with response length a, the intensity $\tilde{s}(a)$ satisfying

$$\tilde{s}(a) = s(x_1) = s(x_2),$$

$x_2 - x_1 = a$ is uniquely determined (see Figure 15). By using the function $\tilde{s}$ which is monotonically decreasing, (5.1) reduces to

$$W_1(a) = w_2 g(a - h_2) + h_1 - \tilde{s}(a). \tag{5.2}$$

Therefore, we see (Figure 16) that there is a stable solution $a_1$. When $s(x)$ is strong, $\tilde{s}(a)$ for small a will be large. The unstable solution $a_0$ will disappear if $h_1 < \tilde{s}(a)$. The excited region of this solution includes the position of the maximum intensity of the stimulus and is of length $a_1$. We see that $a_1$ satisfies

$$\tilde{s}(a_1) = w_2(a_1 - h_2) + h_1 - W_1(a_1).$$

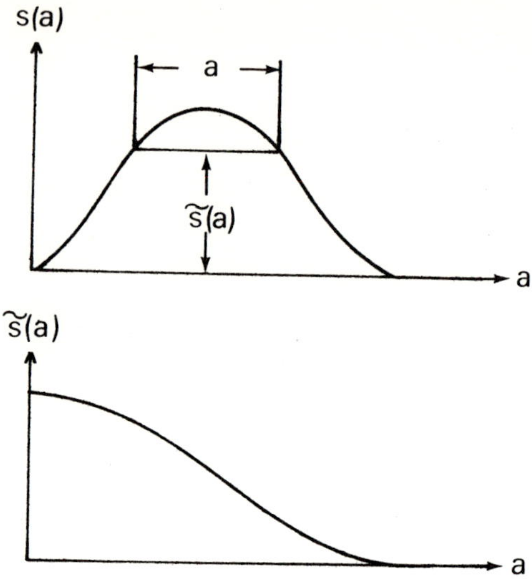


*Figure 15* Relation $\tilde{s}(a)$ between the excitation length and stimulus intensity.

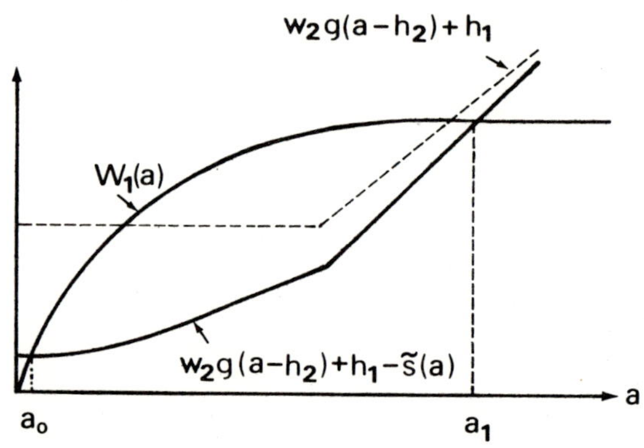


*Figure 16* Graphical solution of equation (5.2).

When the input stimulation includes noise such as shown in Figure 17, the relation $\tilde{s}(a)$ is not monotonic. When a is smaller than the lengths of noisy peaks, it is not unique. It includes small noises of similar frequencies. The equation (5.2) then has many solutions, as is shown in Figure 18. The solutions are distributed around the solutions $a_0$ and $a_1$ for the smoothed input stimulus distribution.

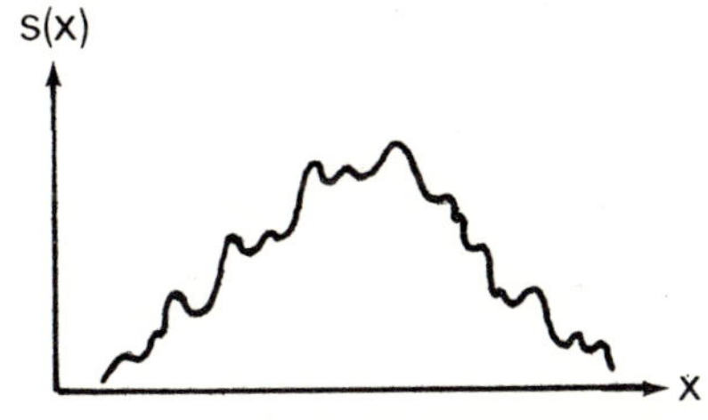


*Figure 17* Noisy input stimulus.

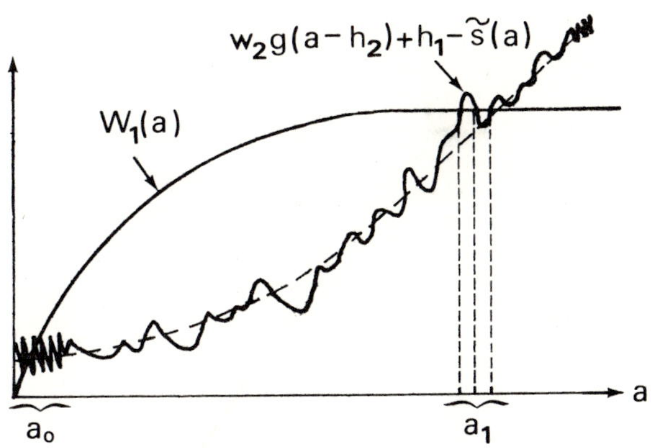


*Figure 18* Graphical solution of equation (5.2) for a noisy input.

Given a distribution of stimuli, the field will converge to one of these equilibrium states (Figure 19). Given a stimulation, the position of the maximum stimulus becomes excited first if its intensity is bigger than the threshold $h_1$. The excitation spreads rapidly to the surroundings because of the excitatory connections of the field. At the same time, it triggers the inhibitory neuron pool. Therefore, the region of excitation, once spread out, begins to reduce by virtue of the effect of inhibition and converges to the equilibrium excitation having the maximum length. Even when the input stimulus is very sharp (like a pulse), it can

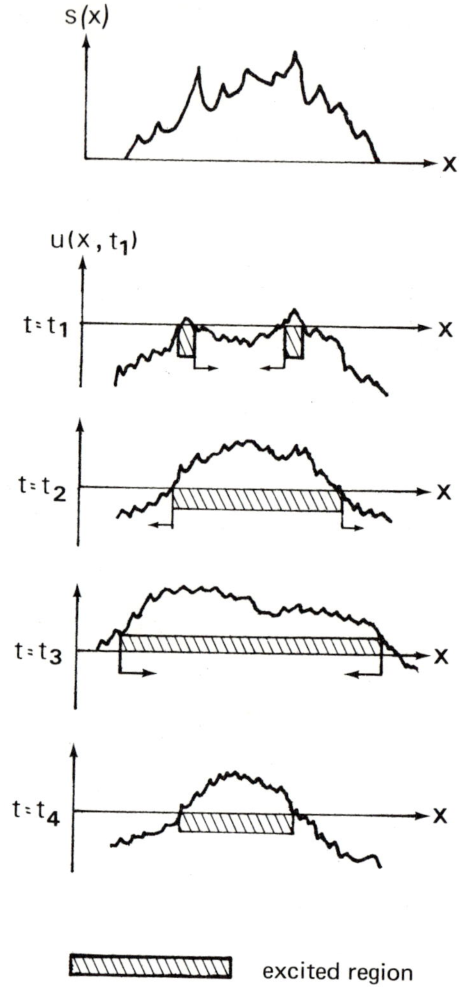


*Figure 19* Dynamical process of excitation in the competition model.

trigger the excitation of the field because the spreading process of the excitation takes place first, the length of the excitation becoming bigger than the critical length $a_0$ (the length of unstable equilibrium) before the inhibitory process is activated.

We now analyze the behavior of the full model when a single-peaked stimulus is applied to a single layer. Replacing (4.2) by (4.1) in the above discussion (monolayer behavior), we deduce that the excited region occurs around the peak of the stimulus, and the length $x_2 - x_1 = a$ of the region satisfies

$$W(a) = h_1 - \tilde{s}(a), \tag{5.3}$$

with the boundary points $x_1$ and $x_2$ satisfying

$$s(x_1) = s(x_2) = \tilde{s}(a).$$

We show an example of the graphs of $W(a)$ and $h_1 - \tilde{s}(a)$. If $\tilde{s}(a)$ is not too large for small a, the equation has two solutions, $a_0$ and $a_1$, as shown in Figure 20. It can be shown that the larger $a_1$ is stable and the smaller $a_0$ is unstable.

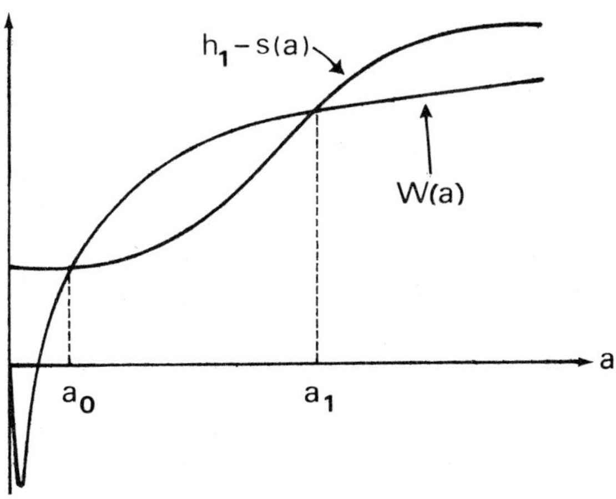


*Figure 20* Graphical solution of equation (5.3).

When the stimulus is weak or when it is not sufficiently wide, the equation has no solution and no excitation occurs. This shows that some *critical coherence length* exists and no excitation occurs unless the stimulus is wider than that length. The critical length depends on the intensity of stimulation.

In order to get the critical coherence length, let us consider uniform stimulation of length a and intensity s. Then, $\tilde{s}(a)$ has the same shape (Figure 21) as s(x). Therefore, given an intensity

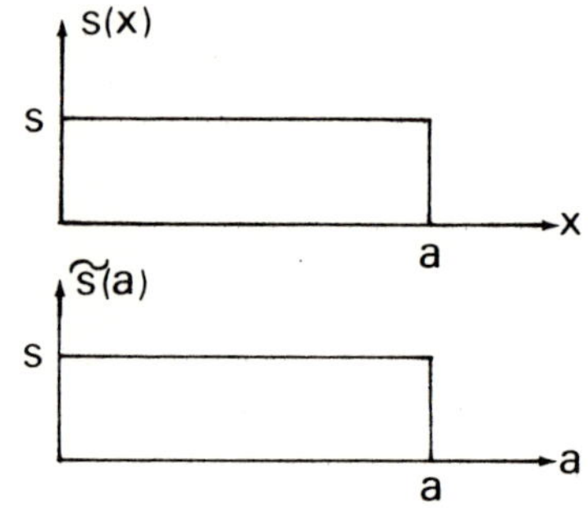


*Figure 21* $\tilde{s}(a)$ for uniform stimulation.

s, we have the critical coherence length a(s) (Figure 22) which is the solution of

$$W(a) = h_1 - s.$$

We can write

$$a(s) = W^{-1}(h_1 - s).$$

It should be noted that, by controlling the threshold $h_1$ of this layer, the critical coherent length can be changed.

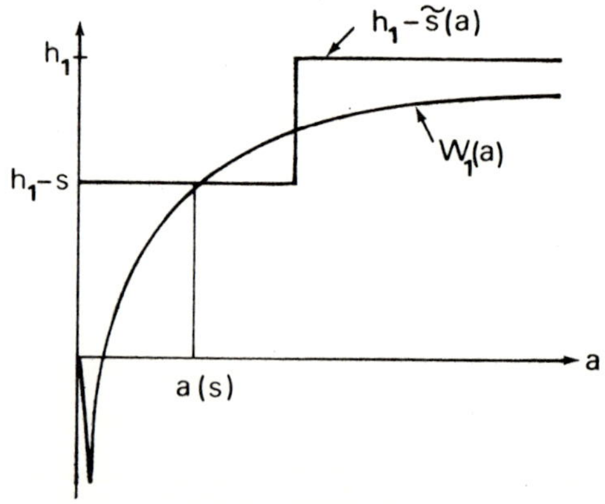


*Figure 22* Critical coherence length for exciting the full model.

In the case of a single-peaked, noisy stimulation, the dynamical process is almost the same as that of the competition field. Since the excitation process comes first, the excited region first grows widely. Then, the inhibitory process becomes active and the excited region reduces. Eventually, it converges to the solution of the maximum possible length.

It should be noted that in Dev's stereopsis model the input stimulation $s_d(x)$ is not a smooth function but consists of a series of pulses showing the match of input from both eyes at the given disparity d. Since the layer has the effect of smoothing this input because neighboring elements are strongly coupled, we may treat a smoothed input function, obtaining the same result. An isolated pulse cannot cause an excitation because it is shorter than the critical coherence length.

## 6. *THE COMPETITION PROCESS*

We can now study competition between multiple peaks. We first examine the competition model and then turn to interaction between the layers of the full model.

### 6.1 *Interaction Between Peaks in the Competition Model*

So far, we have treated a single-peaked, noisy input and seen how the field "searches" for the position of maximum stimulation. The essential feature of the competition model, however, is the competition processes among many peaks of the stimuli. We first show the interaction between two excited regions.

When two excited regions start within a distance smaller than $\delta$ (the half-width of $w_1$--recall Figure 11), they interact directly through the excitatory connections. Since the parts of each excited region close to the other excited region receive stronger stimulation than the outside parts (because the intensity of the connection weight decreases with distance), they attract mutually to unite into one excited region.

When two excited regions exist with a distance bigger than $\delta$, there is no direct interaction between them except for that through the inhibitory channel. Consider the case (Figure 23)

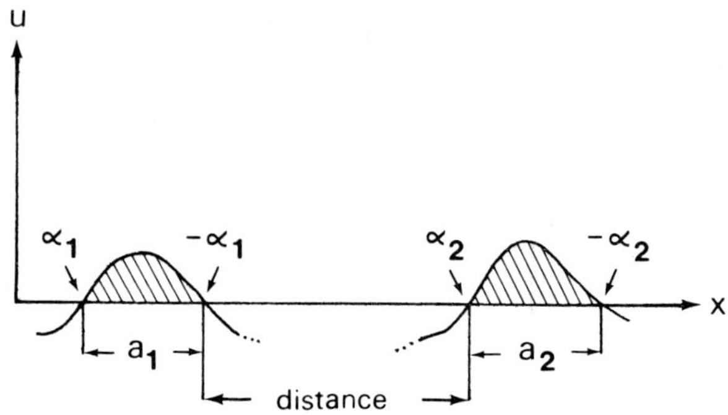


*Figure 23* Interaction between two excited regions.

when two symmetrical, excited regions of lengths $a_1$ and $a_2$ exist, both receiving constant stimulus s and separated by a distance bigger than $\delta$. They neither attract nor repel. Their interaction is represented by the change in their lengths. We adopt (4.5) to write the equations governing the lengths as

$$\dot{a}_1 = \frac{2}{\alpha_1} [W_1(a_1) - w_2 v - h_1 + s],$$

$$\dot{a}_2 = \frac{2}{\alpha_2} [W_1(a_2) - w_2 v - h_1 + s],$$

$$\tau\dot{v} = - v + (a_1 + a_2) - h_2,$$

where $v > 0$ and the $\alpha_i$'s are the gradients of the waveshapes at the boundaries. The equation has two equilibrium solutions satisfying

$$a_1 = a_2,$$

given by solving

$$W_1(a) = w_2 v + h_1 - s = w_2(2a - h_2) + h_1 - s$$
$$= 2w_2 a + (h_1 - h_2 - s).$$

However, both of them are unstable. We show this for the

simplified case where $\tau$ is negligibly small so that v may be equated to $(a_1 + a_2) - h_2$.* The equation reduces, in this case, to

$$\dot{a}_1 = \frac{2}{\alpha_1} [W_1(a_1) - w_2(a_1 + a_2 - h_2) - h_1 + s],$$

$$\dot{a}_2 = \frac{2}{\alpha_2} [W_1(a_2) - w_2(a_1 + a_2 - h_2) - h_1 + s],$$

so that its linearized variational equation has the following coefficient matrix:

$$P = 2 \begin{bmatrix} \frac{1}{\alpha_1}(w_1(a_1) - w_2) & -\frac{1}{\alpha_1} w_2 \\ -\frac{1}{\alpha_2} w_2 & \frac{1}{\alpha_2}(w_1(a_2) - w_2) \end{bmatrix}.$$

When $a_1 = a_2 = a'$, the determinant of P is

$$|\det P| = \frac{4}{\alpha_1 \alpha_2} \cdot [(w_2 - w_1(a')^2) - w_2^2],$$

which is negative (if we put $w_1(a') < 2w_2$--a condition that may fail if a' is very small), and the equilibrium is unstable.

This shows that when two equal regions are excited they compete with each other until the stronger one extinguishes the other. This is true in the case of many excited regions.

We leave open the *rigorous* analysis of the behavior of the unequal regions and the response of the model to unequal sustained inputs with multiple peaks. However, we now informally describe the dynamical process of the field. Given a distribution of input stimuli s(x), the places corresponding to peaks of the input stimuli begin to be activated if the peaks are bigger than the threshold $h_1$ (Figure 24). While the excited regions are growing, the inhibitory pool is activated so that the regions

*It is not difficult but needs a tedious calculation to prove the instability in the general case.

next start to reduce. They compete until one region wins, and the excitation remains at the position of the maximum stimuli belonging to the maximum peak.

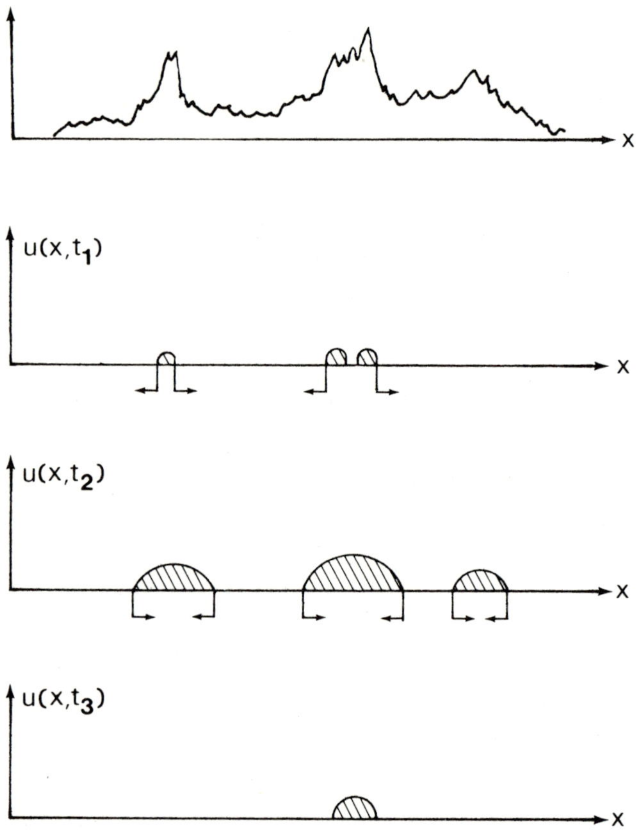


*Figure 24* Dynamical process of the competition model for multi-peaked input.

Now we should mention some speciál phenomena associated with the model. When the input has two (or more) peaks of similar intensity, it may occur that both of the excited regions, once ignited, again disappear by the mutual inhibitory interaction. If the stimuli still remain with the same intensity, one of the regions eventually wins to become excited, suppressing the other.

When two peaks of similar intensity are located within a certain distance, an interesting phenomenon occurs. The places of the stimuli first become excited and then grow (Figure 25). If they are not located too far apart, they are united into one excited region before the inhibitory pool is activated. Then inhibition takes effect, reducing the united excited region. When the intensity of two peaks is similar, the final excitation appears in the middle--the "average fly" effect observed by Ingle. If one is considerably bigger, the reducing velocity of the boundary is much slower for that side of the boundary. Hence, in this case, the excitation region finally comes exactly at the place of the bigger stimulus.

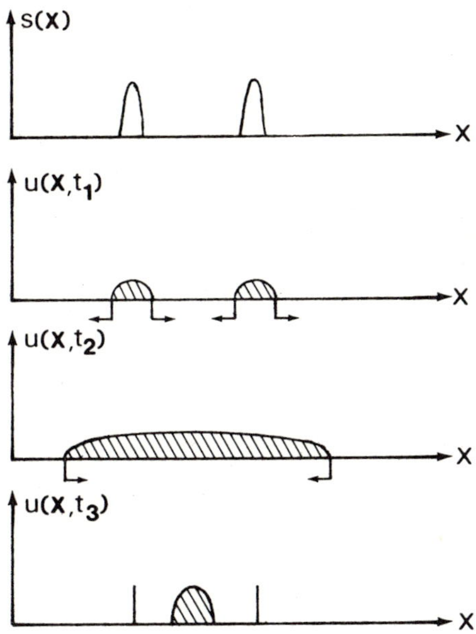


*Figure 25* Response to a double-peaked stimulus.

*6.2 Competition Between Layers in the Full Model*

We now study competition *between* layers in the full model. Let us consider the case where the stimuli, given in layers 1 and 2, are overlapping in part. Let $a_1$ and $a_2$ be, respectively, the lengths of the excited regions of layers 1 and 2, and let r be the length of the overlapped regions of excitation. We first establish a condition which guarantees that this overlap is small. We then analyze this small overlap in more detail. Throughout this analysis, we impose the condition that the maximal stimulus satisfies

$$s_{max} < W_2 + h_1 - W_\infty,$$

where

$$W_2 = \int_{-\infty}^{\infty} w_2(\xi)d\xi.$$

*Theorem 6:* It is impossible for two or more layers of the full model to be excited at the same spatial positions, except for a small overlap [less than the effective length of inhibitory connections $w_2(x)$] at the boundaries of excited regions.

*Proof:* Assume that the excited regions of two layers overlap as shown in Figure 26. Then, the velocity of the boundary $x_2$ of layer 2 can be written as

$$\frac{dx_2}{dt} = \frac{1}{\alpha}[W(a_2) - h_1 + s_2 - w_2],$$

where

$$\alpha = -\frac{\partial u_2(x_2,t)}{\partial x} > 0$$

and $a_2$ is the length of the excited region of this layer. The condition guarantees

$$\frac{dx_2}{dt} < 0,$$

even if layer 2 receives maximum stimulation and $a_2$ is large. The same argument holds for the boundary of layer 1. This shows that there cannot **exist** overlapping of excited regions among two or more layers when the condition holds. □

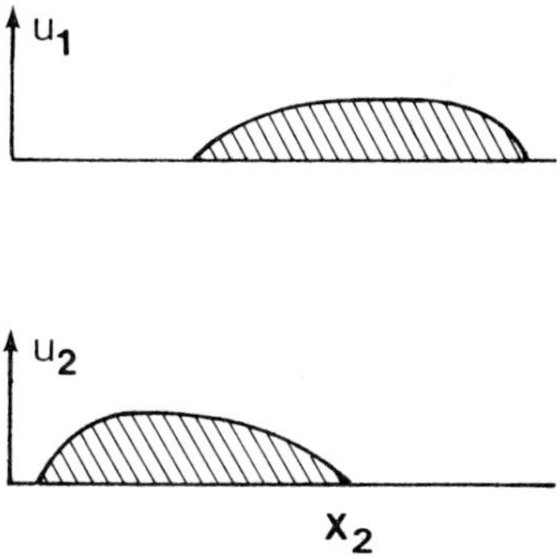


*Figure 26* Overlapping excitation in the full model.

We now study this small overlap in more detail. Let the excited region of layer 1 at time t be from $x_1^1$ to $x_2^1$ and that of layer 2 be from $x_1^2$ to $x_2^2$. We assume, for simplicity's sake, that the outside boundary points $x_1^1$ and $x_2^2$ are fixed and consider how $x_2^1$ and $x_1^2$, i.e., the overlapped region of excitation, changes (Figure 27). We have

$$a_1 = x_2^1 - x_1^1, \qquad \alpha_1 = -\frac{u_1(x_2^1,t)}{\partial x},$$

$$a_2 = x_2^2 - x_1^2, \qquad \alpha_2 = \frac{u_2(x_1^2,t)}{\partial x},$$

$$r = x_2^1 - x_1^2,$$

and

$$\frac{da_1}{dt} = \frac{dx_2^1}{dt},$$

$$\frac{da_2}{dt} = -\frac{dx_1^2}{dt}.$$

Now we assume that the dynamical process of the inhibitory layer is very fast ($\tau \doteqdot 0$) so that

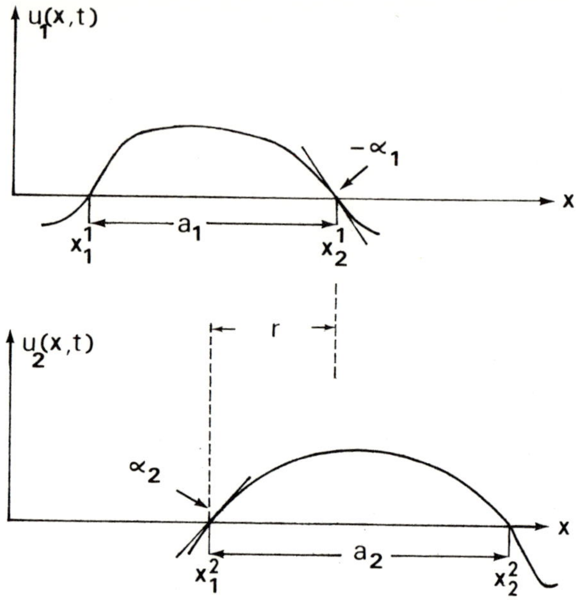


*Figure 27* Dynamics of overlapping stimulation.

$$v(x,t) \doteqdot f[u_1(x,t)] + f[u_2(x,t)] - h_2$$

holds. This means that the activity of the inhibitory layer immediately follows the excitation of the excitatory layers. We have

$$\frac{\partial u_1(x_2^1,t)}{\partial t} = W(a_1) - h_1 + s_1(x_2^1) - J_1,$$

where

$$J_1 = \int_0^r w_2(\xi)d\xi + \int_0^{x_2^2 - x_2^1} (1 - h_2)w_2(\xi)d\xi$$

represents the intensity of inhibition caused by the excitation of the second layer. From

$$\frac{da_1}{dt} = \frac{dx_2^1}{dt} = \frac{1}{\alpha_1}\,\frac{\partial u_1(x_2^1,t)}{\partial t},$$

we have

$$\frac{da_1}{dt} = \frac{1}{\alpha_1}\,[W(a_1) - h_1 + s_1(x_1^1 + a_1) - J_1].$$

Similarly, we have

$$\frac{da_2}{dt} = \frac{1}{\alpha_1}\,[W(a_2) - h_1 + s_2(x_2^2 - a_2) - J_2],$$

where

$$J_2 = \int_0^r w_2(\xi)d\xi + \int_0^{x_1^2 - x_1^1} w_2(\xi)(1 - h_2)d\xi.$$

The equilibrium solution of the excited regions is given by

$$\frac{da_1}{dt} = \frac{da_2}{dt} = 0.$$

At equilibrium, the length of the overlapping excited regions is nearly equal to 0; $r \doteqdot 0$ because of the condition in the previous theorem. We therefore have

$$J_1 = J_2 = \int_0^r w_2(\xi)d\xi + W_2(1 - h_2)$$

around the equilibrium (i.e., when r is small), where

$$W_2 = \int_0^\infty w_2(\xi)d\xi,$$

since $w_2(x)$ is very narrowly spread so that

$$\int_0^{x_2^2 - x_2^1} w_2(\xi)d\xi = \int_0^{x_1^2 - x_1^1} w_2(\xi)d\xi = \int_0^\infty w_2(\xi)d\xi.$$

The equilibrium, hence, occurs at the lengths $a_1$ and $a_2$, satisfying

$$W(a_1) + s_1(x_1^1 + a_1) = W(a_2) + s_2(x_2^2 - a_2).$$

In the case of mutually overlapping input stimuli, the equation has a solution around the center of the overlapping stimuli, and it may have other solutions near the boundaries of the stimulated regions (Figure 28).

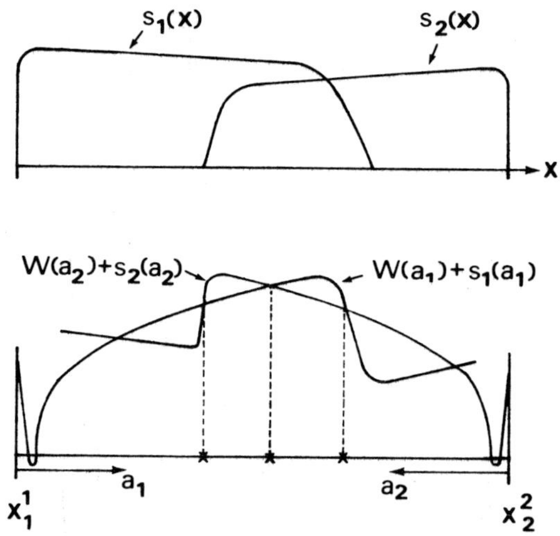


*Figure 28* Equilibrium solutions for overlapping stimulation.

It is interesting to see, given overlapping stimulation, whether the boundary of the excited regions comes at the center or if it comes at the edge of the overlapping stimulation, only one of the two layers getting the excitation over the whole overlapping region. To this end, we study the stability of the equilibrium at the center. The matrix P of the linearized equation for small deviations around the equilibrium is given by

$$P = \frac{1}{\alpha}\begin{bmatrix} w_1(a_1) + s_1'(x_1^1 + a_1) - w_2(r) & -w_2(r) \\ -w_2(r) & w_1(a_2) + s_2'(x_2^2 - a_2) - w_2(r) \end{bmatrix},$$

where we assumed, for simplicity's sake, that $\alpha_1 = \alpha_2 = \alpha$ and $\alpha$ does not depend on $a_1$ and $a_2$. We then have the following theorem:

*Theorem 7:* Put

$$k_1 = w_1(a_1) + s_1'(x_1^1 + a_1),$$

$$k_2 = w_1(a_2) + s_2'(x_2^2 - a_2),$$

where ' denotes the derivative. The equilibrium is unstable when $k_1 < 0$ and $k_2 < 0$, and stable when $k_1 > 0$ and $k_2 > 0$.

*Proof:* We have

$$\mathrm{tr}P = \frac{1}{\alpha}[k_1 + k_2 - 2w_2(r)],$$

$$\det P = \frac{1}{\alpha^2}[k_1k_2 - (k_1 + k_2)w_2(r)].$$

When $k_1 < 0$, $k_2 < 0$ holds, $\mathrm{tr}P < 0$, $\det P > 0$ holds. This shows that the equilibrium is stable. On the contrary, suppose the equilibrium is stable and $k_1 > 0$, $k_2 > 0$ holds. From $\mathrm{tr}P < 0$, $\det P > 0$, we easily have

$$2w_2(r) < h_1 < w_2(r),$$

which is a contradiction (here we take $k_1 > 0$, $k_2 > 0$ into account). Hence, when $k_1 > 0$, $k_2 > 0$, the equilibrium is unstable.

Since $w_1(a_1) > 0$, $w_1(a_2) > 0$, we see that, when uniform stimulations to the two layers are overlapping (Figure 29), $s_1' = s_2' = 0$, and we have $k_1 > 0$, $k_2 > 0$. This shows that one layer overcomes the other on the region of overlapping stimulation. However, if the intensities of stimulation are gradually fading out, as shown in Figure 30, we have $k_1 < 0$, $k_2 < 0$, so that the boundary of the excited regions comes at the middle of the overlapping stimulation. □

Now we can spell out the dynamical processes of the model. Given a stimulus consisting of pulses, the loci of pulses begin to be activated, and then they activate their surroundings. In a region in which pulses are densely packed, the excitation spreads into the whole region. However, in a region in which matching

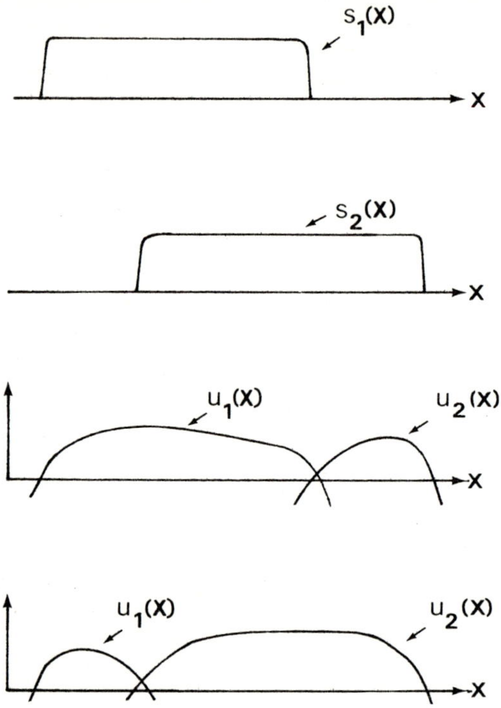


*Figure 29* Two stable equilibria for uniformly overlapping stimuli.

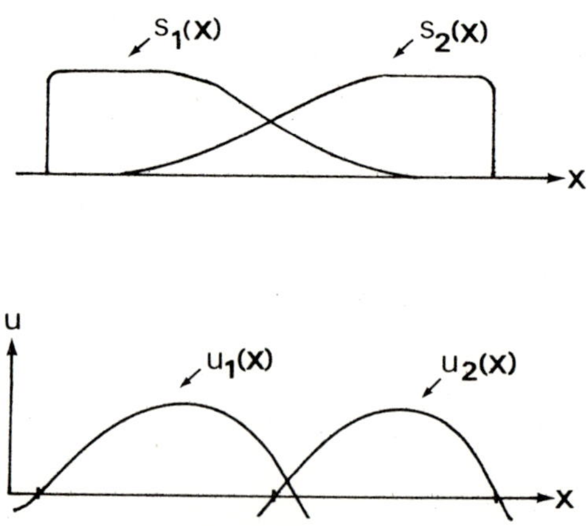


*Figure 30* Equilibrium for gradually decreasing overlapping.

pulses are sparse, the aroused excitation cannot spread but remains solitary. The inhibitory layer then begins to be activated, and it extinguishes these narrow solitary excitations.

When two layers receive overlapping dense stimulation, both layers are excited on the respective stimulated regions, and then the boundaries begin to reduce (Figure 31). The velocity of the reduction depends on the intensity of stimulation. The boundaries at last meet at some point in the overlapping region. If this solution is unstable, the boundaries then move together to one of the edges of the overlapping stimulated regions. Usually, they move in the direction of weaker stimulation. But there is a hysteresis phenomenon, and if the region of the layer of weaker stimulation wins by chance (due to noise), it also gives a stable solution.

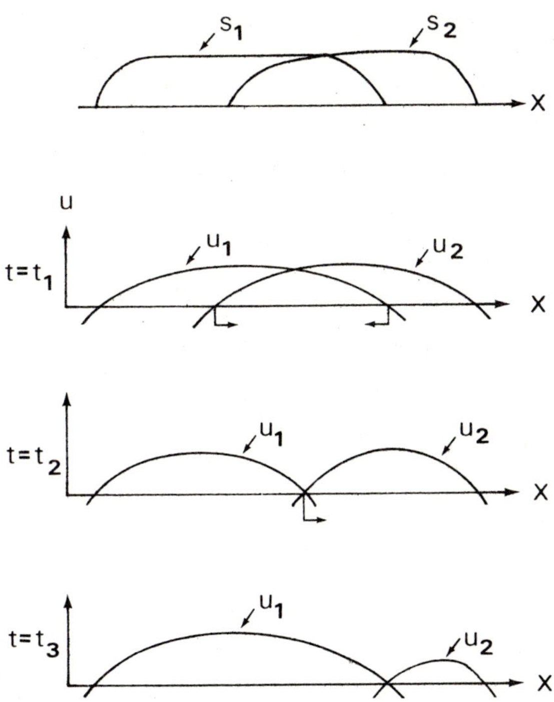


*Figure 31* Dynamical process for overlapping stimulation.

Extending the considerations of the primitive cooperative model of Section 3, we see that the stereopsis model must have a threshold controlling mechanism. By controlling the threshold value of each layer, the model would then be able to jump from one stable solution to another. Multistability and related hysteresis phenomena (with threshold control mechanism) play an important role in the recognition of ambiguous stereograms.

ACKNOWLEDGMENTS *This research was supported in part by a grant from the Alfred P. Sloan Foundation to the Center for Systems Neuroscience, and in part by NIH under Grant No. 5 R01 NS09755 COM of NINCDS.*

## *REFERENCES*

Amari, S. 1971. Characteristics of randomly connected threshold element networks and network systems. *Proceedings of the IEEE 59:* 35-47.

Amari, S. 1972. Characteristics of random nets of analog neuron-like elements. *IEEE Transactions on Systems, Man and Cybernetics SMC-2:* 643-657.

Amari, S. 1974. A method of statistical neurodynamics. *Kybernetik 14:* 201-215.

Amari, S. 1977. Dynamics of pattern formation in lateral-inhibition type neural fields (in preparation).

Arbib, M. A. 1975. Artificial intelligence and brain theory: Unities and diversities. *Annals of Biomedical Engineering 3:* 238-274.

Arbib, M. A., Boylls, C. C., and Dev, P. 1974. Neural models of spatial perception and the control of movement. In W. D. Keidel, W. Händler, and M. Spreng, eds., *Kybernetik und Bionik/Cybernetics and Bionics*. München/Wien: Oldenbourg.

Barlow, H. B., Blakemore, C., and Pettigrew, J. D. 1967. The neural mechanism of binocular depth discrimination. *Journal of Physiology 193:* 327-342.

Dev, P. 1975. Computer simulation of a dynamic visual perception model. *International Journal of Man-Machine Studies 7:* 511-528.

Didday, R. L. 1970. The simulation and modelling of distributed information processing in the frog visual system. Ph.D. Thesis, Stanford University.

Didday, R. L. 1976. A model of visuomotor mechanisms in the frog optic tectum. *Mathematical Biosciences 30:* 169-180.

Ingle, D. 1968. Visual releasers of prey-catching behavior in frogs and toads. *Brain, Behavior and Evolution 1:* 500-518.

Julesz, B. 1971. *Foundations of Cyclopean Perception*. Chicago: University of Chicago Press.

Kilmer, W. L., McCulloch, W. S., and Blum, J. 1969. A model of the vertebrate central command system. *International Journal of Man-Machine Studies 1:* 279-309.

Lettvin, J. Y., Maturana, H., McCulloch, W. S., and Pitts, W. H. 1959. What the frog's eye tells the frog's brain. *Proceedings of the IRE 47:* 1940-1951.

Montalvo, F. S. 1975. Consensus vs. competition in neural networks. *International Journal of Man-Machine Studies 7:* 333-346.

Nelson, J. I. 1975. Globality and stereoscopic fusion in binocular vision. *Journal of Theoretical Biology 49(1):* 1.

Pettigrew, J. D., Nikara, T., and Bishop, P. O. 1968. Binocular interaction on single units in cat striate cortex. *Experimental Brain Research 6:* 391-410.

Scheibel, M. E., and Scheibel, A. B. 1968. The brainstem core--An interactive matrix. In M. Mesarovic, ed., *Systems Theory and Biology*. Berlin/New York: Springer-Verlag.

# The Amygdaloid Kindling Phenomenon: A Tentative Model

ISRAEL LIEBLICH*
*Center for Systems Neuroscience*
*University of Massachusetts at Amherst*
SHUN-ICHI AMARI†
*Center for Systems Neuroscience*
*University of Massachusetts at Amherst*

ABSTRACT *A tentative model which accounts for the strongest phenomena defining kindling is suggested. It is based on an excitatory-inhibitory coupling of neural aggregates, to which a self-stimulating element for the excitatory aggregate was added.*

*The functional linking hypothesis views the representation of kindling as a process of gradual transition through structural changes from a stable system, to a system showing stability for small perturbations and an oscillatory orbit for larger perturbations, to a purely oscillatory system.*

*The anatomical linking hypothesis views the excitatory aggregate as representing the hypothalamus, the inhibitory aggregate as representing the hippocampal-septal-preoptic complex, and the self-stimulating element of the excitatory aggregate as representing the amygdaloid-pyriform complex. The model was realized on a digital computer with graphic capabilities and showed good qualitative agreement with the experimental data related to kindling. In addition, the use of the model for generating new experiments is discussed.*

---

*1975-76 Fellow, on leave from the Department of Psychology, The Hebrew University of Jerusalem, Jerusalem, Israel.

†1975-76 Fellow, on leave from the Department of Mathematical Engineering and Instrumentation Physics, University of Tokyo, Tokyo, Japan.

## 1. *INTRODUCTION*

The gradual development of permanent change in patterns of brain electrical activity and the appearance of behavioral automatisms and convulsions as responses to brief electrical stimulation, due to daily such stimulation of various limbic sites, has been termed by Goddard, McIntyre, and Leech (1969)--"kindling". Goddard et al. also provided most of the parametric data related to this phenomenon. It is the point of the present paper to provide a first approximation model which accounts for some of the strongest phenomena defining kindling, and to derive from it some experimental predictions which might further clarify this striking phenomenon.

Experimental studies of the kindling phenomenon may choose to emphasize the changes in electrical activity aspects observed from the kindled and adjacent sites (e.g., Racine, Newberry, and Burnham, 1975), the changes in behavioral aspects of kindling (e.g., Pinel, Phillips, and McNeill, 1973; Lieblich, Yitzhaki, and Cohen, 1977), or both aspects (e.g., Goddard et al., 1969). While there is not any fixed relation between *first* EEG changes and *first* behavioral manifestations of kindling, the electrical and behavioral aspects of the phenomenon are so intimately linked that each of them has served as an excellent definition of it.

In the present study, we have chosen to provide a model that would produce, in principle, the EEG changes described during the development of kindling. It should be emphasized that choosing the EEG changes for modeling does not in any way make them more important or interesting than the behavioral aspects of kindling. They were chosen only because we could, at this stage, define a "simple" linking hypothesis (Weisstein, Ozog, and Szoc, 1975) between the model and the experimental data by using the EEG changes observed in kindling.

The kindling phenomenon is defined in EEG terms, using the amygdaloid complex as a kindling target. It is a change from practically no EEG response to electrical stimulation of the kindled area at the initial stage of the experiment, to a relatively sustained afterdischarge response of high-amplitude, slow-rhythmic spikes recorded from the kindled site. This change occurs in the amygdala after about seven daily stimulations. These afterdischarges spread spatially to other limbic regions (Goddard et al., 1969; Racine et al., 1975).

### *1.1 The Functional Linking Hypothesis*

We look for a system, the behavior of which can be described by a set of differential equations. The system should display stability under normal conditions, having a unique equilibrium. When stimulated occasionally, the system returns to equilibrium and stays at it. (More realistically, the system should fluctuate around the equilibrium as a result of random inputs from its surroundings.) The repetitive electrical stimulation results in some permanent changes in the system's parameters, such as threshold or connectivity values. These changes, as they develop, induce the system to display oscillatory behavior. While the equilibrium of the system stays stable, a stable oscillatory orbit appears around it. Still, at this stage any noise influence may prevent the permanent maintenance of the oscillatory orbit, and the system returns to equilibrium. As kindling proceeds, the equilibrium becomes purely oscillatory. The oscillation persists in this case without limit.

The observation that instigated the present model is related to the relative ease with which it was possible to induce kindling from different structures. The following is a list of the main regions found by Goddard et al. (1969) to be responsive to the kindling procedure, the numbers in brackets representing the mean number of trials to first convulsion in a given region:

1. Amygdala [15],
2. Globus pallidus [22],
3. Anterior limbic field [29],
4. Olfactory area [29],

5. Enthorhinal cortex [37],
6. Olfactory bulb [44],
7. Septal area [55].
8. Preoptic area [63],
9. Caudate-putamen [74], and
10. Hippocampus [77].

The following functional generalization presents itself: It is easier to get kindling from behaviorally excitatory systems than from behaviorally inhibitory systems. The proposed model demonstrates how a specific interaction between inhibitory and excitatory regions may be sufficient for kindling to manifest itself.

### *1.2 Structural Considerations and the Anatomical Linking Hypothesis*

Figure 1 describes the suggested structural model. It can be shown that any coupling of an inhibitory and excitatory system by

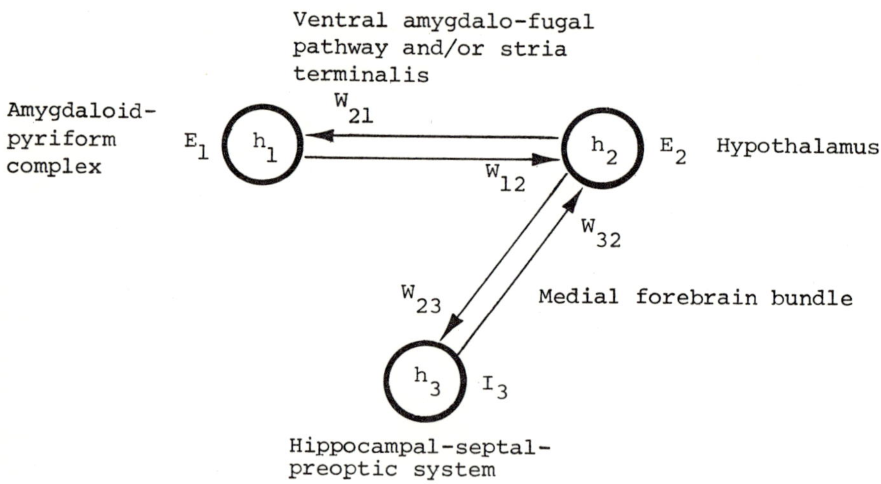


*Figure 1* Schematic representation of the model for kindling. $E_i$ and $I_i$ = neural aggregates. The connection from $I_3$ to $E_2$ is inhibitory. $h_i$ = threshold values of the aggregates. $W_{ij}$ = connectivity strength values from i to j.

itself would make it very difficult to fulfill the necessary essential characteristics of kindling. Thus, an inhibitory-excitatory coupling only, such as $I_3$ and $E_2$, would not produce the desired result. It was felt that adding a self-excitation element $E_1$ to $E_2$ would be sufficient to produce the desired change.

It should be emphasized that there are no empirical grounds for any claim that the anatomical representation in Figure 1 or in any other configuration is *unique*. However, it was felt that the representation in Figure 1 is an attractive possibility. Since this study is a first approximation, it is hoped that proper experimentation might easily refute or modify the suggested representation. The point of the model is that the three interacting anatomical components are sufficient to produce a change from virtual stability into subsequent oscillations around the stable equilibrium.

It may be recognized that the anatomical configuration chosen is an attempt to integrate the knowledge of major descending afferent and ascending efferent connections of the hypothalamus (Nauta and Haymaker, 1969) with the mapping of kindling from limbic sites (Goddard et al., 1969). It is of interest to note that in the latter work, the values for the hypothalamus proper do not appear. It is known, however, that stimulation of the amygdaloid complex produces afterdischarges in the hypothalamus. It is also clear from studies within the context of self-stimulation that it is possible to trigger convulsions from positive self-stimulation sites by use of higher current intensities. It is also possible to elicit motor automatisms (Koella, 1969) observed in kindling and stimulus-bound behaviors (Valenstein, Cox, and Kakolewski, 1970) by appropriate stimulation of hypothalamic sites. Thus, the attempted incorporation of the hypothalamus as a component related to the kindling phenomenon may be promising in the future experimental test of the model.

## 2. FORMAL DESCRIPTION OF THE MODEL

The model is composed of three nerve tissue aggregates, two of which, $E_1$ and $E_2$, are purely excitatory and the third, $I_3$, which is purely inhibitory. The behaviors of the three aggregates are described by the following differential equations:

$$\dot{E}_1(t) = -\frac{1}{\tau_1}(E_1 - h_1) + W_{21}f_2[E_2(t)],$$

$$\dot{E}_2(t) = -\frac{1}{\tau_2}(E_2 - h_2) + W_{12}f_1[E_1(t)] - W_{32}f_3[I_3(t)],$$

$$\dot{I}_3(t) = -\frac{1}{\tau_3}(I_3 - h_3) + W_{23}f_2[E_2(t)] + W_{13}f_1[E_1(t)],$$

where

$E_i(t)$, $i = 1,2$, and $I_3(t)$ are the total membrane potentials of the aggregates;

"." denotes time differentiation;

$\tau_i$, $i = 1,2,3$, is the time constant of the potential decay;

$W_{ij}$, $i = 1,2,3$, $j = 1,2,3$, represents the strength of connections between the aggregates;

$h_i$, $i = 1,2,3$, is the threshold value of the aggregates, which includes effects of constant stimulations and inhibitions from outside of the system; and

$f_i$, $i = 1,2,3$, is the nonlinear function which relates the membrane potential of an aggregate to some output (e.g., firing rate, observed EEG, etc.).

Two generalizations from the anatomical and experimental data have been used to constrain the model:

1. The ascending hypothalamic efferents to the structures involved in the model are less marked than the descending afferents from these structures. In terms of the model, this means

$$W_{12} > W_{21},$$

$$W_{32} > W_{23}.$$

In addition, we decided to set $W_{13} = 0$ since we did not feel that this connection is substantial anatomically, and it would not help to clarify the behavior of the model at this stage.

2. Goddard et al., (1969) reported that kindling was virtually impossible from tracts, and that kindling when appearing was confined to nucleic sites. In terms of the present model, we assumed that the changes undertaken to produce kindling would be limited to changes in the threshold of the aggregates and not to changes in connectivity strength between aggregates.

### *3. COMPUTER REALIZATION OF THE MODEL*

The model was realized by using a modification of a phase space (e.g., Rabenstein, 1966) plotting program.* It was developed and run on a GT-44 graphics computer system, which is based on a DEC PDP-11/40 processor. For its input, the program asks the user for the following information:

1. Time constant values for the three aggregates;
2. Connectivity values between the aggregates;
3. Threshold values for the three aggregates; and
4. Steepness values, $S_i$, for the nonlinear functions as associated with the outputs of the three aggregates.

The form of the nonlinear functions is

$$f_i(u) = \frac{1}{\pi} \tan^{-1} \left( \frac{u}{S_i} - 1 \right) + \frac{1}{2},$$

where

$i = 1,2,3$, represents the aggregate;

$S_i$ represents the steepness of $f_i$ of u - the membrane potential of the aggregate; and

$f_i$ varies asymptotically between 0.0 and 1.0.

---

*The original phase space plotting program was written by Fred K. Lenherr of the Center for Systems Neuroscience, University of Massachusetts at Amherst. We wish to thank Arthur Karshmer for his help in the modification of the graphics of the plotting program.

Since the behavior of the model could not be described analytically, the following heuristics were used to approximate the desired effects:

1. Time-constant values were arbitrarily chosen to be 10 for both $E_1$ and $E_2$, and 20 for $I_3$, preserving a convention that inhibitory central processes show a larger time constant than excitatory processes.

2. Steepness values for nonlinearity of the different aggregates were chosen intuitively and fixed at 0.7 for both $E_1$ and $E_2$, and at 1.0 for $I_3$.

3. Under these conditions, connectivity values were chosen (a) to satisfy the above mentioned anatomical inequalities, (b) such that around the equilibrium points the system is stable but has oscillatory tendencies. The values of connectivities chosen after some trial and error were:

$$W_{12} = 4.0,$$
$$W_{21} = 2.0,$$
$$W_{23} = 3.1,$$
$$W_{32} = 5.0, \text{ and}$$
$$W_{13} = 0.0.$$

4. Threshold values were chosen such that at equilibrium, the potential output of the aggregates would be nearly 0.0. The following values result in stability of the system:

$$h_1 = -5.2,$$
$$h_2 = 2.5, \text{ and}$$
$$h_3 = -15.5.$$

## 4. *RESULTS*

As defined in the Introduction, we consider the kindling process to be represented by a transition from a stable state of the system to a state in which we observe local stability for sufficiently small perturbations and a limit cycle for stronger perturbations. The clearest way to view the behavior of the above described system is through plotting the trajectory of the system.

*4.1 Effects of Changes in Threshold of $E_1$ (Amygdaloid-Pyriform Complex)*

Figures 2, 3, and 4 show examples of a two-dimensional projection of the trajectory of the system on the $E_1$, $E_2$ plane. As may be seen in Figure 2, at the threshold value of $E_1$ of -5.2, the system is stable. A reduction of threshold of $E_1$ to -5.17 results in stability around the origin and in a limit cycle otherwise (Figure 3). A reduction of threshold to -4.8 results in a purely oscillatory system (Figure 4).

The system is very sensitive to the changes in threshold of $E_1$. Increases in the thresholds of $E_1$ result in stabilization of the system. It is of interest to note that for extreme reductions of the threshold of $E_1$, stability is once again observed. At this stage, we would rather limit the maximum reduction of the threshold of $E_1$ to -4.8.

*4.2 Effects of Changes in Threshold of $E_2$ (Hypothalamus)*

It is impossible to get kindling in the defined sense through reduction of the threshold (stimulation) of $E_2$, at least in the ranges at which we tested the system. Going into a kindled state by reduction of threshold of $E_1$ can be reserved into a stable state by strongly stimulating $E_2$ (reducing its threshold). With further reductions in threshold of $E_1$, the system may be kindled, once more achieving criterion at about -5.1. Enhancement of the threshold of $E_2$ results in the hastening of the kindling process.

*4.3 Effects of Changes in Threshold of $I_3$ (Hippocampal-Septal-Preoptic System)*

Stimulation of $I_3$ stabilizes the kindled system at the critical threshold of $E_1$ = -5.17. But by further lowering the threshold of $E_1$ to -5.0, the system will go into the kindled state. Elevation of the threshold of $I_3$ has interesting effects. By moderate changes it is possible to kindle the system, but extreme

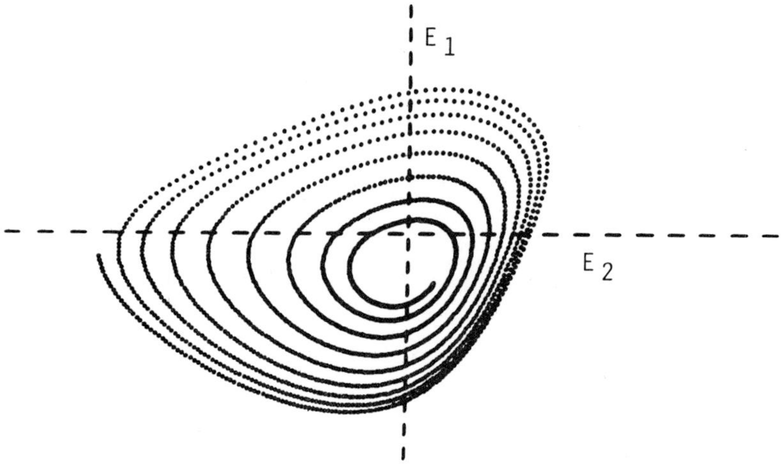


*Figure 2* Two-dimensional projection of the trajectory of the system in the stable state. $E_1$ = amygdaloid-pyriform aggregate. $E_2$ = hypothalamic aggregate. Maximum values on both axes = 3.0. Threshold of $E_1$ = -5.2. Development in time: counterclockwise.

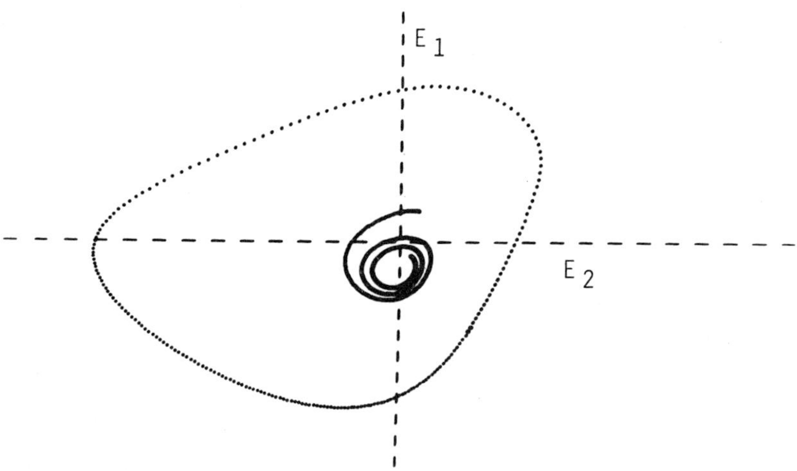


*Figure 3* Two-dimensional projection of the trajectory of the system near fully developed kindling. Note stability for small perturbation and a limit cycle for stronger stimulation. The symbols and maximum values as in Figure 2. Threshold of $E_1$ = -5.17. Development in time: counterclockwise.

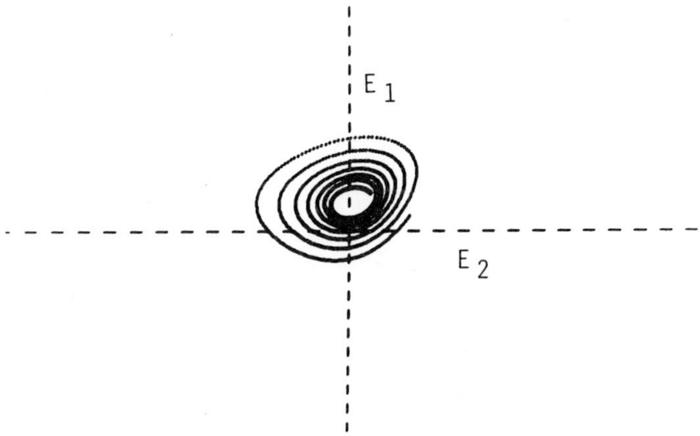


*Figure 4* Two-dimensional projection of the trajectory of the system in an extreme fully oscillatory state. The symbols and maximum values as in Figure 2. Threshold of $E_1$ = -4.8. Development in time: counterclockwise.

enhancement of thresholds result in stabilization of the system. No kindling is possible at very high thresholds of $I_3$.

*4.4 Effects of Changes in Thresholds of Both $E_2$ and $I_3$*

Stimulation (reduction in threshold) of $E_2$ and $I_3$ stabilizes the kindled system at threshold $E_1$ = -5.17. The system goes to a kindled state, however, by further reduction of threshold of $E_1$ to -4.5. Table 1 summarizes the main results.

## 5. *DISCUSSION*

The dynamic characteristics of neuron aggregates have attracted considerable attention. Amari (1971) has shown that discrete time systems consisting of excitatory and inhibitory neuron aggregates may display oscillatory behaviors and has proven the stability condition. These results were also proven for continuous time systems (Amari, 1972; Wilson and Cowan, 1972). Feldman and Cowan (1975) attempted to apply neural oscillation behaviors in a system consisting of six aggregates to the brainstem. Ahn and Freeman (1974) showed that a system consisting of two

*TABLE 1*

Summary of some of the effects of threshold changes of the different components of the system on kindling.

| *Experiments* | *THRESHOLD* *Enhancement of* $E_1$ | $E_2$ | $I_3$ | *Reduction of* $E_1$ | $E_2$ | $I_3$ | *Results* |
|---|---|---|---|---|---|---|---|
| 1 | | | | X | | | kindling* |
| 2 | X | | | | | | stabilization* |
| 3 | | | | | X | | stabilization |
| 4 | | X | | | | | kindling |
| 5 | | | | | | X | stabilization |
| 6 | | | X | | | | kindling |
| 7 | | | | | X | X | stabilization |

*At threshold $E_1$ = -5.17.

aggregates, each of which is a second-order system, may also display oscillations, and applied it to the olfactory complex.

However, all the above mentioned oscillations are simple in the sense that the stable oscillatory orbit does not coexist with the stable equilibrium. In order to get such a new type of oscillatory system, it was necessary to add a self-stimulation component ($E_1$) to the excitatory ($E_2$)-inhibitory ($I_3$) coupling. It is clear from the results of the present study that our present system displays a stable oscillation around a stable equilibrium. This behavior, which characterizes the kindled state, is derived from the stable system through structural change.

Inspection of the results of the model seems to indicate good qualitative agreements with the known phenomena of kindling. Most remarkable to us is some clarification of the puzzling effects of transient arousal on seizure behavior. The model shows interference with the expression of a "kindled" situation by stimulation of the hypothalamic ($E_2$) aggregate. This is in line with a wealth of findings which showed that phasic arousal may interfere with convulsive and seizure phenomena in general (e.g., Arduini and Lairy-Bounes, 1952; Lieblich and Guttman, 1965; Roelofs, 1967; Chorover and DeLuca, 1969) and with kindling (Pinel et al., 1973; Lieblich et al., 1977).

The model predicts that the process of kindling (or perhaps any development of convulsive behavior) may be retarded by direct stimulation (or otherwise reducing the threshold) of the hypothalamus. In addition, the model shows that kindling would not be possible from the hypothalamus even with quite drastic reduction in its thresholds.

On the other hand, a reduction in stimulation (an elevation of the threshold) of the hypothalamus would accelerate kindling. Roelofs (1967) demonstrated the interaction of the effects of initial behavioral state and focus of brain stimulation on seizure susceptibility. It was found that amygdaloid seizure susceptibility is significantly increased in drowsy and sleep states, and is decreased in wakeful and active states.

The model points to an extreme dependence of the kindling phenomena on the threshold of the amygdaloid aggregate. This is in line with Goddard et al. (1969) who described the stabilizing effect of a lesion at the electrode tip implanted in a kindled amygdala.

The complexity of the experimental possibilities, with regard to the kindling phenomenon, may be exemplified by the following: Given the above cited experimental results, which indicate that phasic arousal may retard convulsive behavior, what could be expected by stimulation of a behaviorally inhibitory system? The model has a prediction. Under the chosen parameters, stimulation of $I_3$ stabilizes the kindled system. What then may be expected after elevation of threshold of the behaviorally inhibitory system? The answer of the model is that, under the chosen parameters, moderate elevations in the threshold of $I_3$ result in facilitation of kindling, *but* extreme elevation of the threshold of the inhibitory component stabilizes the kindled system. It would be extremely difficult to kindle such a system in the first place by changing the thresholds of $E_1$ and $E_2$.

Thus, the suggested model clearly exhibits the paradoxical effects which tranquilizing and energizing agents may have on behavior. As a matter of fact, the model, after further tests, could serve for testing suggested procedures for treatments of seizure behavior. In addition, the model would suggest extreme caution in the interpretation of changes in electrical responses to stimuli recorded from limbic structures as a function of lesions placed in hypothesized tracts.

It should be remembered, however, that the model does not address itself to the way in which threshold changes in the neural aggregates are achieved. We conceive of them informally as resulting from changes in synaptic characteristics following any stimulus applied systematically to an interconnected excitatory or inhibitory aggregate. Kindling observed from a behaviorally inhibitory area is viewed as representing an elevation of threshold through self-inhibition, while kindling observed from an excitatory aggregate is brought about by reduction of threshold through self-excitation.

The model, as a first approximation, does not handle possible field effects, but is rather a pointwise analysis of the three aggregates. Also, we did not address ourselves to the exact EEG characteristics observed during kindling. In our view, this should be postponed until more confidence in the model is built up. At this stage, enough testable predictions using the model may be generated so as to easily reject or modify it.

*REFERENCES*

Ahn, S. M., and Freeman, W. J. 1974. Steady state and limit cycle activity of a mass of neurons forming simple feedback loops. I. Lumped circuit model. *Kybernetik 16:* 87-91.

Amari, S. 1971. Characteristics of randomly connected threshold element networks and network systems. *Proceedings of the IEEE 59:* 35-47.

Amari, S. 1972. Characteristics of random nets of analog neuron-like elements. *IEEE Transactions of Systems, Man and Cybernetics, SMC-2:* 643-657.

Arduini, A., and Lairy-Bounes, G. C. 1952. Action de la stimulation electrique de la formation reticulaire du bulbe et des stimulations sensorielles sur les ondes electriques. *EEG and Clinical Neurophysiology 4:* 503-512.

Chorover, S. L., and DeLuca, A. M. 1969. Transient changes in electrographic reaction to ECS in the rat following foot-shock. *Journal of Comparative Physiological Psychology 69:* 141-149.

Feldman, J. L., and Cowan, J. D. 1975. Large-scale activity in neural nets. II. A model for brainstem respiratory oscillator. *Biological Cybernetics 17:* 39-51.

Goddard, G. V., McIntyre, D. C., and Leech, C. K. 1969. A permanent change in brain function resulting from daily electrical stimulation. *Experimental Neurology 25:* 295-330.

Koella, W. P. 1969. Control of skeletal motor activity, with emphasis on the role of diencephalic mechanisms. In W. Haymaker, E. Anderson, and W. J. H. Nauta, eds., *The Hypothalamus*. Springfield: Charles C Thomas.

Lieblich, I., and Guttman, R. 1965. The relation between motor activity and risk of death in audiogenic seizures of DBA mice. *Life Sciences 4:* 2295-2299.

Lieblich, I., Yitzhaki, J., and Cohen, E. 1977. Effects of septal lesions on behavior elicited by stimulation of the amygdaloid complex. *Behavioral Biology* (in press).

Nauta, W. J. H., and Haymaker, W. 1969. Hypothalamic nuclei and fiber connections. In W. Haymaker, E. Anderson, and W. J. H. Nauta, eds., *The Hypothalamus*. Springfield: Charles C Thomas.

Pinel, J. P. J., Phillips, A. G., and McNeill, B. 1973. Blockage of highly stable kindled seizures in rats by antecedant foot-shock. *Epilepsia 14:* 29.

Rabenstein, A. L. 1966. *Introduction to Ordinary Differential Equations*. New York: Academic Press.

Racine, D., Newberry, F., and Burnham, W. M. 1975. Post-activation potentiation and the kindling phenomenon. *EEG and Clinical Neurophysiology 39:* 261-271.

Roelofs, G. A. 1967. Behavioral state and seizure susceptibility in cats: A study of responses to electrical stimulation of the basal temporal lobe in cats under different behavioral conditions. *Acta Neural Scandinavia 43(26)*.

Valenstein, E. S., Cox, V. C., and Kakolewski, J. W. 1970. Reexamination of the role of the hypothalamus in motivation. *Psychological Review 77:* 16-31.

Weisstein, N., Ozog, G., and Szoc, R. 1975. A comparison and elaboration of two models of metacontrast. *Psychological Review 82:* 325-343.

Wilson, H. R., and Cowan, J. D. 1972. Excitatory and inhibitory interactions in local populations of model neurons. *Biophysical Journal 12:* 1-24.

# Two Time-Domain Oscillatory Brain Models for Psychologists

WILLIAM L. KILMER
*Department of Computer and Information Science*
*and*
*Center for Systems Neuroscience*

ISRAEL LIEBLICH*
*Center for Systems Neuroscience*
*University of Massachusetts at Amherst*

THOMAS H. PROBERT
*Department of Computer and Information Science*
*University of Massachusetts at Amherst*

ABSTRACT *Attention is drawn to several phenomena related to kindling, α blocking, massed and spaced learning, and self-stimulation. Two time-domain models and their computer simulations are presented. These models introduce depletion and resonant frequency concepts. The experimental phenomena are discussed in terms of the two models.*

## 1. *INTRODUCTION*

Much of the current theoretical thinking in psychology is centered around probabilistic-combinatorical considerations. One possible explanation of this phenomenon may be related to the serious training in the use of statistical inference tools for the interpretation of experiments, which may result in intimate

---

*1975-76 Fellow, on leave from the Department of Psychology, The Hebrew University of Jerusalem, Jerusalem, Israel.

knowledge of probability theory and combinatorics. This training bias may result in models in these terms.

It is of interest to note that when exceptions to the above described bias are found, some of the most impressive models to date in psychology present themselves (e.g., Weisstein, Ozog, and Szoc, 1975, model for metacontrast).

We would like to further draw attention in this theoretical note to some important phenomena in psychology for which a time-domain approach might be promising. Given the importance of the phenomena and the relative straightforwardness of the time-domain application to them, it is hoped that more attention will be invested in the further education of psychologists to time-domain concepts.

Consider the following set of experimental observations:

1. Goddard, McIntyre, and Leech (1969) reported that the low-level electrical stimulation at 10 cps applied to limbic structures is optimal for triggering learned "kindled" convulsions. Stimuli below 10 cps and above 300 cps are practically noneffective in eliciting an otherwise very stable convulsive response.

2. The same authors report that, in the stage of learning the "kindled" convulsive response, the intertrial interval between the electrical stimuli applied to the limbic structures involved in the phenomenon is of critical importance. Interstimulus intervals of less than five minutes resulted in lower convulsion rates than interstimulus intervals greater than 30 minutes.

3. Gallistel (1973), following Deutsch (1964), has shown that if a rat is given widely spaced trials in a runway with rewarding intracranial stimulation (RIS) primings at the start and reinforcing stimulations at the goal, then the influence of priming declined when stimulation pulse-pair intervals were shorter than 1 msec. The reinforcing value of the stimulation applied at the goal fell only when the pulse-pair intervals were less than 0.6 msec.

4. Segal and Bloom (1974) showed that hippocampal pyramidal cells were inhibited by locus coeruleus stimulation. With ten pulses per second, and after five seconds of stimulation, the cells remain inhibited for as long as two minutes after the end of stimulation.

5. Massed versus spaced learning has been an issue for a long time in learning theory. McGaugh, Westbrook, and Burt (1961) have shown that differences between maze-bright rats and maze-dull rats are reflected in their sensitivity to massing of learning trials. While learning in widely spaced conditions, no difference may be detected between learning capabilities of the two strains. Only under massed trials do the maze-bright animals show superiority. McGaugh et al. also showed that the use of picrotoxin may counteract the effects of trial massing in maze-dull rats.

6. The most potent variable to influence the probability of observing $\alpha$ rhythm in human subjects (Ss) is the state of their eyes. In a given S, the probability of observing $\alpha$ rhythm is much higher given closed eyes than when S's eyes are open (Lippold, 1973).

7. Lieblich, Isseroff, Kleiner, and Cohen (1976) have observed clear oscillations in bar pressing rates in a dyadic paradigm in which rats delivered RIS to each other, but could not deliver it to themselves.

It is clear that each isolated phenomenon of the type presented above may be explained by itself, using some "minitheory" that would fit only the specific set of observations which served to suggest the theory in the first place.

We posit that all the above-mentioned phenomena may have one common mechanism. All of these may point to a mechanism of depletion in the neuronal matrix controlling the various phenomena. The following theoretical formulation is an attempt to describe the possible components of a depletion/regeneration process, and to point qualitatively to the ways in which it could be used to generate experiments that might support, refute, or modify it.

We shall next present two general models, and then discuss the above phenomena in terms of them.

## 2. *FIRST MODEL*

We begin by reinterpreting the well-known logistic equation of population biology (Kilmer and Probert, 1976),

$$\frac{dN(t)}{dt} = rN(t) \left[1 - \frac{N(t)}{K}\right], \tag{2.1}$$

so that N(t), instead of denoting number of individuals in a population, denotes the normalized aggregate firing rate of a population of neurons. Because N(t) must always be positive or zero, dN(t)/dt, the rate of change of N(t), henceforth denoted $\dot{N}(t)$, is zero only if

$$\left[1 - \frac{N(t)}{K}\right] = 0. \tag{2.2}$$

In (2.1), r is a constant and K is a variable that sets the steady state value of N in (2.2). Hence, K reflects the effective excitatory input to the neuronal circuit in question. Note that if $N(t) < K$, $\dot{N} > 0$; and if $N(t) > K$, $\dot{N}(t) < 0$. Thus, N(t) always approaches K.

We now introduce a new term into (2.1) to represent the fractional depletion D(t) of neurotransmitter and supporting substances in our neuronal circuit at t as a function of past and present N values. We do this by simply multiplying K by [1 - D(t)], where D(t) is constrained to lie between 0 and 1. Model One is

$$\dot{N}(t) = rN(t) \left[1 - \frac{N(t)}{K[1 - D(t)]}\right], \tag{2.3}$$

together with the convolution integral [to produce a weighted summing from (t - T) to t]

$$D(t) = \int_{t-T}^{t} \frac{FN(\tau)}{[1 - D(\tau)]} P(t - \tau)d\tau. \tag{2.4}$$

In (2.4), $F[N(\tau)]/[1 - D(\tau)]$ represents the rate of neural depletion caused by N at time $\tau$ on the undepleted resource fraction $1 - D(\tau)$; and $P(t - \tau)$ is the fraction of the depletion incurred at time $\tau$ that still persists at t. The F and P functions used in a computer solution of (2.3) and (2.4) are shown in Figure 1.

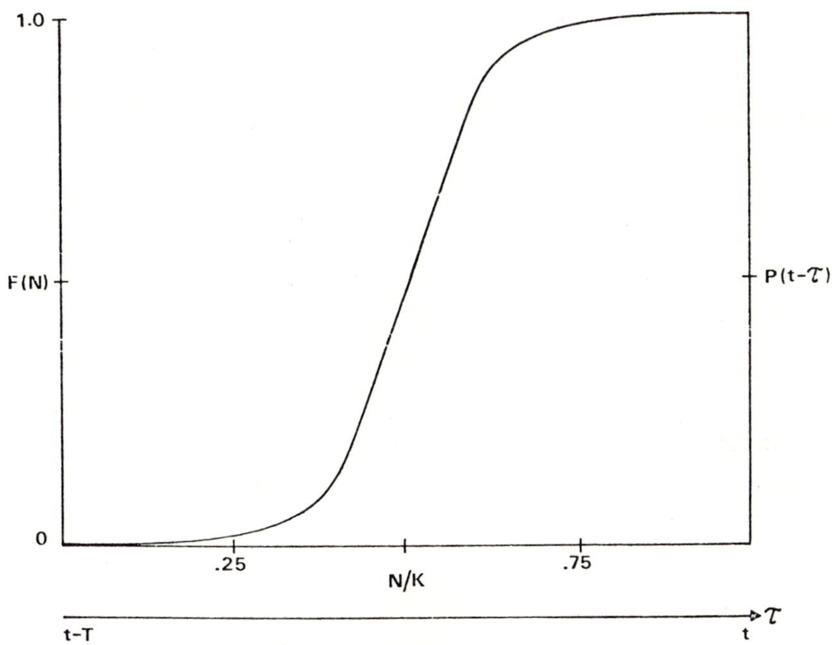


*Figure 1* The F and P functions of equation (2.4).

Note that in (2.3), $\dot{N} = 0$ when $N = K(1 - D)$. In (2.4), D increases with increasing N, but D changes much more slowly than N if P extends over a period that is long compared to N's (1/r) rise constant. The precise shapes of the curves in Figure 1 appear not to matter.

Figure 2 shows how N and D vary with time in a computer solution of (2.3) and (2.4). Other computer runs confirmed that the shape of the N curve becomes increasingly like that defined for N by (2.1) as r decreases relative to T. As r increases, N and D become more oscillatory with longer cycle times.

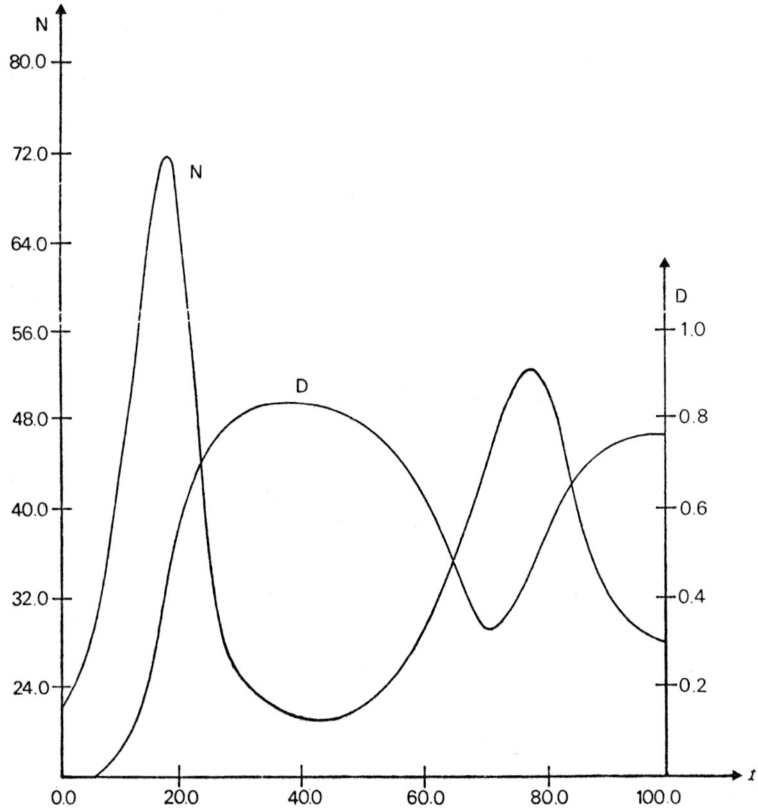


*Figure 2* Initial segments of N and D versus time in a simulation of equations (2.3) and (2.4).

### 3. *SECOND MODEL*

Model Two consists of (2.3) and a different D(t) equation. Writing the rate of change of D(t) at t as $\dot{D}(t)$, and using the set of constants $C_i > 0$ to relatively weight the various terms, our D(t) equation is

$$D(t) = \frac{\varepsilon W^2}{K}\left[F_t + F_\Delta + F_S\right], \tag{3.1}$$

where

$$F_t = N(t) + C_1 N(t)[1 - D(t)]^2, \tag{3.1a}$$

$$F_{\Delta} = - C_2N(t - \Delta) - C_3N(t - \Delta)\ [1 - D(t - \Delta)], \tag{3.1b}$$

$$F_S = - C_4K \int_{C_0}^{t} [D(t) - C_5]dt, \tag{3.1c}$$

subject to $0 < D(t) < 1$.

In (3.1a), the first term of $F_t$ represents the present rate of neural depletion with a basic term $N(t)$ and an adjustment term $N(t)\ [1 - D(t)]^2$. The adjustment fraction $[1 - D(t)]^2$ reflects a decreased depletion rate at higher $D(t)$ values where neural transmission strengths fall off increasingly fast--faster than linearly, but probably slower than cubicly--hence, the square on $[1 - D(t)]$. In (3.1b), $F_\Delta$ represents the normal renewal rate of the neural depletion incurred at time $(t - \Delta)$. Thus, as a first approximation, we are assuming that normal recovery from depletion requires $\Delta$ time units. The first term in $F_\Delta$ is supplemented by a $[1 - D(t - \Delta)]$-diminished $N(t - \Delta)$ term to take into account the falloff in normal renewal due to the $N(t - \Delta)$-overloading of the $D(t - \Delta)$-depleted neuronal circuit at $t - \Delta$. Such overloading presumably draws on neural reserves that have to build up again before the associated $D(t - \Delta)$ can be fully removed. This more complex process is accounted for by $F_S$.

To interpret $F_S$, we need to know that, subject to conditions to be imposed below on r in (2.3) and the $C_i$ in (3.1), N and D possess stable oscillatory solutions as shown in Figure 3 for $\varepsilon = 1$. There, $F_S$ is negative in the A intervals and positive in the B intervals, and $\Delta$ is the period of oscillation. $F_S$ represents the regenerative or reductive influence on $D(t)$ associated with the N and D values over the preceding T time units, $T \le \Delta/2$. During the A intervals, where $F_S$ is negative, $F_S$ reduces D to reflect recent low N values, which together with the high D condition always lead into a restorative phase for the neural substances in question. During the B intervals, positive $F_S$ values represent the increased depletion resulting from precursor and endproduct decreases caused by the preceding high N values.

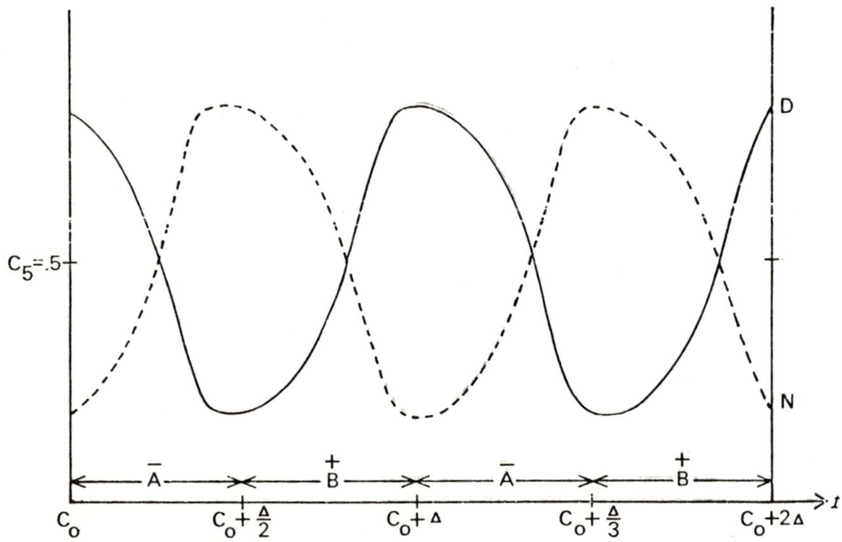


*Figure 3* N and D of equation (3.1) versus time. $\varepsilon = 1$, $\Delta$ = one oscillatory period, and the average $F_S$ is $C_5 = .5$.

We have shown elsewhere (Kilmer and Probert, 1976) that (3.1) can be transformed into the famed Van der Pol (1926) oscillator equation,

$$\frac{d^2x}{dt^2} + \varepsilon[x^2 - 1]\frac{dx}{dt} + x = 0, \tag{3.2}$$

for which $-W \le x \le W$, with the W here the same as in (3.1). The transformation required the following assumptions: $C_1 = 4/3$, $C_2 = 1/W^2$, $C_3 = 2$, $C_4 = 1$, $C_5 = 1/2$, $N(t) = N(t - \Delta)$, $D(t) = D(t - \Delta)$, and r of (2.3) so large relative to $\dot{D}$ that N is always approximately $K[1 - D(t)]$. The generality of the transformation from (3.1) to (3.2) derives from (3.2) being a special case of a large class of extremely robust oscillators defined by

$$\frac{d^2x}{dt^2} + f(x)\,\frac{dx}{dt}\,g(x) + g(x) = \varepsilon(t), \tag{3.3}$$

where $f(x)$, $g(x)$, and $\varepsilon(t)$ are only mildly constrained. [(3.3) is the renowned Liénard equation (Graef, 1972; Grafton, 1972).]

Figure 4 shows solutions to (3.2) for different values of $\varepsilon$, which is the waveshape and period-determining factor in (3.2) (Van der Pol, 1926).

Given our assumption that $N = K(1 - D)$, these solutions define slow (with respect to $1/r$), antiphasic D and N oscillations as suggested in Figure 3. Note that the re-expression of (3.2) given in (3.1) contains the waveshape and frequency parameter $\varepsilon$ on the left side of the equation only. We might, therefore, interpret $\varepsilon$ as an adjustable or trainable adaptation parameter.

## 4. *APPLICATION OF THE MODELS*

We shall now interpret the seven phenomena in the Introduction in terms of the above two models. We suppose that certain regions of neuronal circuitry, because of their anatomical and physiological characteristics, can sustain large amplitude spiking and slow potential activities at certain overall population frequencies. We shall call such frequencies "resonant", and shall associate them not only with seizure phenomena but also with broadcast reinforcement signals and with recognition responses to certain animal situations.

The seven phenomena will now be discussed one after the other:

1. Kindling stimulation between 60 and 300 cps could permanently change the circuit $\varepsilon$ in the equation-(3.1) sense of Model Two. $\varepsilon$ would change to a seizure-resonant frequency setting. The seizure frequency would, in general, be a function of, but not equal to, the stimulation frequency.

2. The interval between kindling sessions might have to be such that at least most sessions occurred at high-N times in the Figure-2 sense of Model One. The low-N intervals in Figure 2

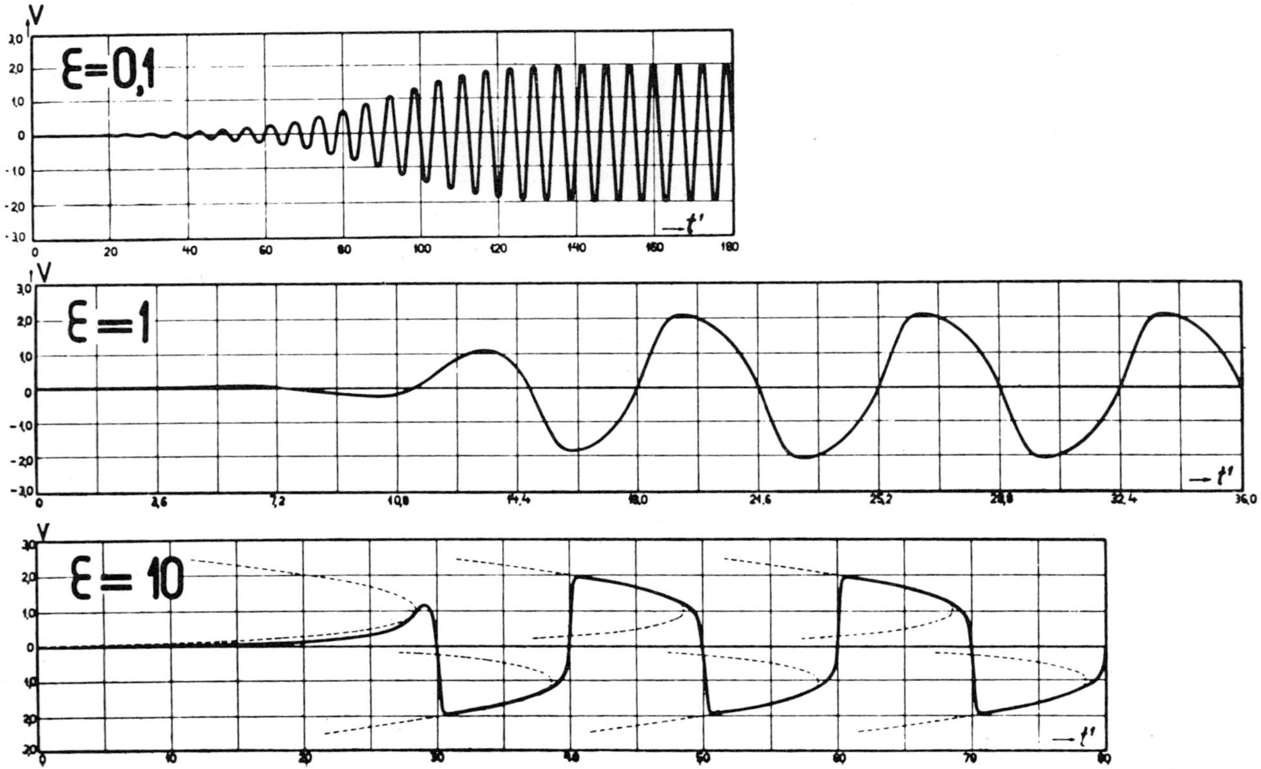


*Figure 4* Solutions of equation (3.2) plotted versus time. (From Van der Pol, 1926. Van der Pol's V is our N.)

would correspond to intervals that would be largely refractory to permanent kindling effects. The depleted substance here would be different than in (1) above because different time scales are involved.

3. Here we invoke the idea of resonant frequencies for broadcast reinforcement signals. RIS signals at the proper frequencies would effect reinforcement in situation-recognizing resonant circuits by permanently changing these circuits' $\varepsilon$'s in the equation-(3.1) sense of Model Two. Such $\varepsilon$ changes, which might be synaptically implemented, would enable recognition circuits to respond with greater amplitude to the presence of priming and goal situations. The new $\varepsilon$'s would thus lead to stronger communication links between the different situation-recognizing and motor circuits. This scheme suggests that a frequency-related, state-dependent type of learning occurs and might explain why the need for priming persists.

4. Ten-cps locus coeruleus stimulation might temporarily change a Model-Two $\varepsilon$ for the inhibitory interneuronal circuitry (IIC) in the hippocampus, moving it into IIC's resonant range. The resonance, once established, might persist for as long as two minutes before dying out. This phenomenon would probably not be regarded as an IIC seizure.

5. Referring to Figure 2 for Model One, massed trials might be ineffective if employed beyond the first high-N peak. Maze-bright rats would then have a longer high-N duration than maze-dull ones. Picrotoxin might increase the persistence time T of depletion in equation (2.4) and thereby effectively change maze-dull rats into maze-bright ones.

6. The excitatory drive stemming from open eyes might drop the visual cortical circuit's $\varepsilon$ from 10 to 0.1 in the Figure-4 sense of Model Two. This would change the circuit's dominant oscillatory frequency from 10 cps to 70 cps. If the circuit was resonant at 10 cps but not at 70 cps, the 70-cps EEG signal would

flatten and phasically scatter. The resulting EEG would contain low voltage, fast activity.

7. We interpret this phenomenon in the Figure-2 sense of Model One. That is, we assume each rat first recognizes its situation and then begins bar pressing. If the rat's situation-recognizing circuit has passed into the low-N range following the first N peak by the time the rat begins bar pressing, RIS would be ineffective. Later, when the rat reached its second N peak, RIS would resume its effectiveness. At this time, the second rat, by bar pressing, would reinforce the first rat.

Other phenomena, including habituation, have been shown to behave in Model-One fashion with respect to signal transmission parameters. Also, a host of Model Two-like phenomena, involving adaptations and adjustments, have been looked at but not studied extensively.

It is hoped that psychologists will explore other experimental paradigms and models to search for new time-domain effects and mechanisms of the general type suggested by equations (2.3), (2.4), and (3.1).

*REFERENCES*

Deutsch, J. A. 1964. Behavioral measurement of the neural refractory period and its application to intracranial self-stimulation. *Journal of Comparative Physiological Psychology 56:* 1-9.

Gallistel, C. R. 1973. Self-stimulation: The neurophysiology of reward and motivation. In J. A. Deutsch, ed., *The Physiological Basis of Memory*. New York: Academic Press.

Goddard, G. V., McIntyre, D. C., and Leech, C. K. 1969. A permanent change in brain function resulting from daily electrical stimulation. *Experimental Neurology 25:* 295-330.

Graef, J. 1972. On the generalized Liénard equation with negative damping. *Journal of Differential Equations 12:* 34-62.

Grafton, R. B. 1972. Periodic solutions of certain Liénard equations with delay. *Journal of Differential Equations 11:* 519-527.

Kilmer, W. L., and Probert, T. 1976. Oscillatory depletion models for renewable ecosystems. *Department of Computer and Information Science Technical Report 76-5*. University of Massachusetts at Amherst.

Lieblich, I., Isseroff, A., Kleiner, M., and Cohen, E. 1976. What may be expected when two brains stimulate electrically each other's lateral hypothalamus. *Brain Theory Newsletter 1:* 44-47.

Lippold, O. 1973. *The Origin of the Alpha Rhythm*. London: Churchill Livingstone.

McGaugh, J. L., Westbrook, W., and Burt, G. 1961. Strain differences in the facilitative effects of 5-7-dipheny 6-1-3-diazamantan 6-01 (1757 1.S) on maze learning. *Journal of Comparative Physiological Psychology 54:* 502-505.

Segal, M., and Bloom, F. E. 1974. The action of norepinephrine in the rat hippocampus. II. Activation of the input pathway. *Brain Research 72:* 99-114.

Van der Pol, B. 1926. Relaxation oscillation. *Philosophical Magazine 2:* 978-992.

Weisstein, N., Ozog, G., and Szoc, R. 1975. A comparison and elaboration of two models of metacontrast. *Psychological Review 82:* 325-343.

# Some Design Criteria for the CORETEX Language

ARTHUR I. KARSHMER
*Department of Computer and Information Science*
*and*
*Center for Systems Neuroscience*
*University of Massachusetts at Amherst*

FRED K. LENHERR
*Center for Systems Neuroscience*
*University of Massachusetts at Amherst*

ABSTRACT *One of the primary research goals of the Center for Systems Neuroscience is the design and development of a highly interactive neural modeling language which makes extensive use of computer graphics techniques. This paper reviews some of the modeling tools currently available to the neural modeler as well as discusses the basic features which should be incorporated in a neural modeling language. To show the advantages of this approach, the FORTRAN encoding of a typical neural model is compared with the much more readable version possible in a CORETEX-like language.*

## 1. *INTRODUCTION*

Despite many years of effort in the computer simulation of neural models, no generally accepted language has as yet evolved for talking about highly parallel neuronal networks. As a result, the investigator must rely on languages designed for other purposes while conducting his simulations.

Although the most general simulation tool is "assembly" language (almost all computer software, for example, is translated into assembly language), it is clear that encoding models at this low level would discourage even the most dedicated researcher as it demands that he be familiar with all of the idiosyncrasies of each particular computer. At the assembly language level, programming even the most simple data manipulations becomes a complex chore.

Higher level languages, such as FORTRAN, offer a much better vehicle for describing neural networks, but FORTRAN, too, suffers from its inability to state problems of a parallel nature in a way that would facilitate the design and implementation of neural models. The basic control structures found in FORTRAN, such as the "DO" loop and "GO TO" statement, force the user to think about his problem in a serial way. This, of course, creates a situation in which the researcher must go through several levels of translation before his model can be implemented.

Other high-level programming languages, such as ALGOL, PASCAL, SNOBOL, and LISP, do give the researcher the ability to define more complex data types which more closely approximate the structures which are needed to simulate the actions of neural networks. However, even these high-level languages do not provide a convenient means of expressing the parallel functions associated with most neural network dynamics. The high-level programming language which seems to offer the researcher the most powerful means of approximating the parallel computations necessary to simulate a neural network is APL (A Programming Language) because of its dynamic nature and built-in array manipulation features.

At this point, a simple example might help us to understand one of the fundamental differences between a serial language, such as FORTRAN, and a pseudo-parallel language, such as APL. In neural modeling it is often necessary to manipulate arrays of data which represent the current state of a given network--so, for our current example, let us assume that we have a 4 x 3 array which contains the state of our model. Let us also assume that the array is called "NEURONS" and contains the following arbitrary data:

$$\text{NEURONS} = \begin{bmatrix} 1 & 2 & 3 \\ 4 & 5 & 6 \\ 7 & 8 & 9 \\ 10 & 11 & 12 \end{bmatrix}$$

In FORTRAN we would have to write the following program segment to get the sum of the four rows and the sum of the entire array:

```
1    IARRAY = 0    !**COUNTER FOR ARRAY
     DO 10 I = 1, 4
     IROW = 0    !**COUNTER FOR ROWS
     DO 20 J = 1, 3
     IARRAY = IARRAY + NEURONS(I, J)    !**ADD TO ARRAY TOTAL
20   IROW = IROW + NEURONS(I, J)    !**ADD TO ROW TOTAL
10   TYPE 30, IROW    !**TYPE ROW TOTAL
     TYPE 30, IARRAY    !**TYPE ARRAY TOTAL
30   FORMAT (I6).
```

This program segment would produce the following output:

```
 6
15
24
33
78.
```

In APL, on the other hand, we could do summation over the rows and the entire array in a much simpler fashion.

The APL statement

```
+/NEURONS
```

would produce a vector which contains the sums of the rows,

```
6    15    24    33,
```

while the APL statement

```
+/+/NEURONS
```

produces the summation of the entire array (which is 78).

In addition to the general purpose programming languages previously described, several general purpose simulation systems are also available for conducting research in neural modeling. Languages such as GASP, GPSS, MIDAS, SIMULA, and ISL, while allowing the researcher to specify pseudo-parallel computations, do not permit him to describe his model in terms which are meaningful to the neuroscience community as a whole. The following brief discussions of GPSS and SIMULA should help to point out why they are not well suited to the needs of the neural modeler.

### *1.1 GPSS*

The General Purpose Systems Simulation, GPSS, is a computer program aimed at simplifying the task of carrying out simulation studies (Blake and Gordon, 1964). In GPSS the user describes the structure of his model in the form of a block diagram which is drawn with a fixed set of predefined block types. Each of these block types is intended to represent a specific action that characterizes some operation that can occur in a system. The actual flow or sequence of actions that do occur in the specific model are indicated by connecting lines between the blocks of the diagram. Several connections can be drawn from a given block where a choice of actions exists.

Moving through the simulated system or model are certain basic units which depend on the nature of the system being simulated. For example, a communication system is concerned with movement of messages, a traffic system with vehicles, or a data processing system with data records. Each of these units is identified with entities called transactions. "The sequence of these actions occuring in the system in real time is reflected in the movement of transactions from block to block in simulated clock time" (Efron and Gordon, 1964).

Each of the block types is represented in GPSS by a particular symbol and is also given a name which describes the actions of the block. Figure 1 shows the names and symbols of several of the block types concerned with creating, delaying, and removing transactions in a GPSS simulation. The block type called ORIGINATE is used to create and enter transactions into the simulation, with the action time controlling the interval between successive creations. To remove a transaction from the simulation, the user simply executes a TERMINATE block. An ADVANCE block is used to represent any action that takes time and delays the transaction, but does not involve the use of any equipment. The ADVANCE block can also be used as a buffer in which transactions can be held while waiting for equipment to become available or for some condition in the system to change.

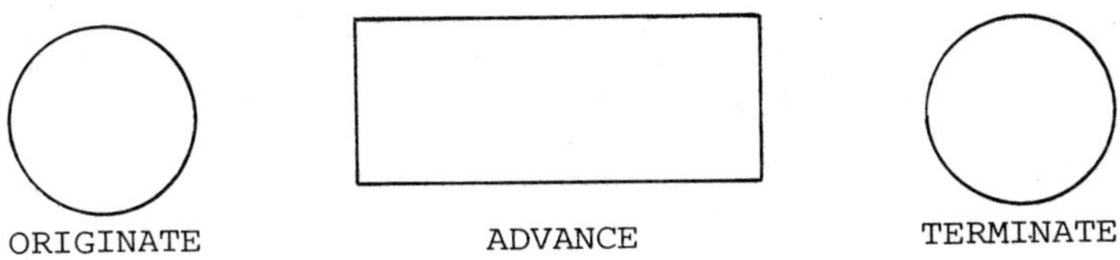


*Figure 1* (After Blake and Gordon, 1964.)

In Figure 2 we show a number of blocks associated with the use of equipment by transactions. The number of the equipment item to be used by the block is indicated in the flag attached to the symbol. A HOLD block permits a transaction to engage a facility for as long as the transaction remains active in the block.

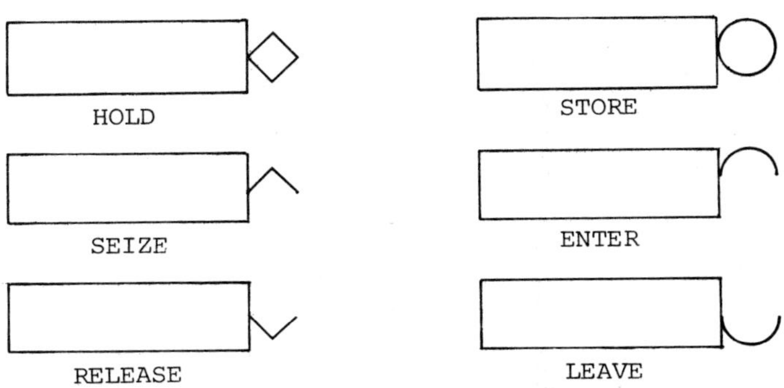


*Figure 2* (After Blake and Gordon, 1964.)

The SEIZE block also permits a transaction to engage a facility, but with the exception that control of the facility is not relinquished until some later time when a RELEASE block is entered. The STORE block allows a transaction to occupy space in a storage while it is in the block, while an ENTER block permits a transaction to occupy space, and a LEAVE block allows space to be released.

GPSS allows transactions to be kept waiting at a block while the conditions for advancing a transaction are not satisfied. Movement of such transactions are kept in order under program control by a FIFO (first in, first out) rule. Information regarding the transaction queue is not kept unless the queue forms a QUEUE block, shown in Figure 3, which is designed to measure the average and maximum queue lengths and, if required, the distribution of time spent in the queue.

The MARK and TABULATE blocks, also shown in Figure 3, are used to measure the length of time taken by transactions moving through the system or parts of the system. The TABULATE block also enters information into a table relating to the difference between the times of arrival at the MARK and TABULATE blocks. The TABULATE block may be used, in addition to working in conjunction with a MARK block, to tabulate a wide variety of different data relating to the system being simulated.

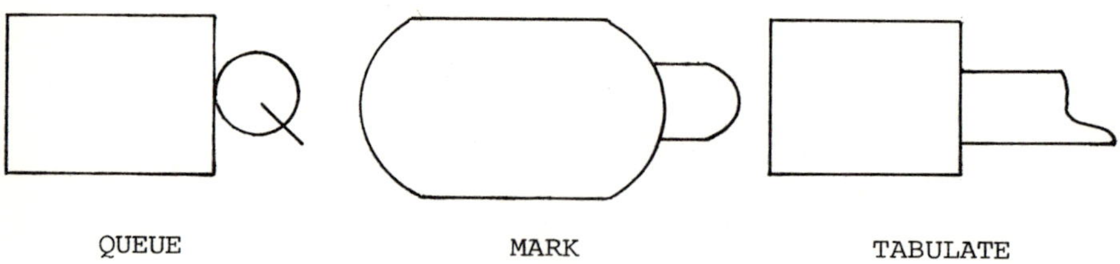


*Figure 3* (After Blake and Gordon, 1964.)

By using the block types available in GPSS, the system simulator or modeler is able to express his model in terms of a block diagram and then map the block diagram on a one-to-one basis into GPSS instructions. While this system is ideal for modeling systems such as traffic flow and message switching, it does not lend itself to the neural modeling environment as it does not offer a rich or meaningful enough set of primitives relating to neural models.

### *1.2 SIMULA*

SIMULA 67 is an extension to the ALGOL 60 programming language which includes special features to aid in the modeling process. The central concept in the SIMULA programming language is that of "CLASS." This concept combines into a single entity the concepts of structure, process, program, and procedure.

The rules of the ALGOL 60 language have been designed in such a way as to ensure that the lifetimes of block instances are nested, in the sense that those instances that are the first to go out of existence are those that are the last to be activated. This particular feature is the one which allows ALGOL 60 implementations to take advantage of the "stack" as a method of dynamic memory allocation. This scheme, while being efficient and straightforward to implement, does have the disadvantage that a program which creates a new block instance can never communicate with it as an object which exists and has attributes since it has disappeared by the time the calling program has regained control (Dahl, Dijkstra, and Hoare, 1972). The calling program then can only observe the results of the execution of the procedures it references.

In SIMULA, on the other hand, a block instance may outlive its calling statement and may remain in existence for as long as the calling program needs to refer to it. Because of this extra complexity, it is no longer possible to manage storage using a simple stack system; instead, more complicated garbage-collection schemes are required to manage memory. This extra level of complexity is acceptable because it permits a much wider range of concepts to be easily expressed. In particular, it gives rise to the "CLASS" concept, which is central to the SIMULA simulation language.

A "CLASS" can be defined as a procedure which is capable of giving rise to block instances which can survive the call which brought them into existence. This concept is a way of partitioning both program code and data, which seems quite natural since data and the operations on them logically belong together. A SIMULA "CLASS" may be declared, with or without parameters, in exactly the same way as a standard ALGOL procedure. The "CLASS" extension to ALGOL permits the modeler to execute transactions in a simulation and communicate with them at any later point in the simulation run with no fear of having them vanish. This one key point is the basis of the SIMULA 67 simulation language.

Besides the "CLASS" concept, SIMULA 67 offers other extensions to the ALGOL language which facilitate the simulation process. Listed below, without further explanation, is a partial list of the high-level simulation functions which are primitives in SIMULA:

RESUME
HOLD
WAIT
ACTIVATE
SIMULATE.

While SIMULA 67 does offer many high-level extensions to ALGOL which are powerful tools in the hands of the systems simulator, it does not begin to offer the primitives desirable for neural modeling. For example, SIMULA 67 does not offer any convenient or easily expressible method for specifying cellular or synaptic dynamics, not to mention complicated patterns of connectivity between cells.

### *1.3 CORETEX*

Even though there are numerous general purpose simulation languages currently available, none were designed to solve the unique problems faced by the neural modeler. In the remainder of this paper, we will attempt to isolate a number of the basic features which should be included in a neural simulation language, as well as discuss some of the issues facing the implementation of such a language. We have chosen to call our proposed language "CORETEX" --a combination of "CORE", suggesting the presence of a computer, and "CORTEX", which suggests its relevance to the study of brain systems.

## *2. DESIRED CAPABILITIES OF THE LANGUAGE*

A crucial first step in the process of designing a language for neural simulation or, for that matter, any programming language, is simply to list some of the most desirable features of such a language. This section lists several, but certainly not all, of the features desired in a truly general purpose neural modeling language.

### *2.1 Fidelity*

The first and probably most important feature of our language is "fidelity"--which in this case simply means that the results produced by a simulation should accurately reflect the intentions of the programmer. While at first glance this condition might seem trivial, it should be kept in mind that programs written in CORETEX will be translated into some other form before being executed by the computer. It is in this mapping from CORETEX language statements to machine instructions that the language designer and implementer must pay particular attention to the problem of fidelity. What this means is that we must be able to guarantee that the results obtained from running a CORETEX program are those which are truly appropriate to the object of study rather than those from a closely related but different neuronal network.

### *2.2 Flexibility*

A second and equally important point to be considered is that the proposed language be as flexible and general as possible as it will be used by a wide variety of neural researchers with many differing orientations. For this reason, the proposed language must be able to simulate complicated multilayered networks as well as simple single-cell systems.

### *2.3 Historian*

The third general capability to be built into the CORETEX language is a means by which the system can save intermediate results for later use. This feature is necessitated by the vast amount of data usually generated by even the simplest neural model. It would be a great convenience to the user if our proposed language provided simple and understandable methods for dealing with this high volume of intermediate data rather than forcing the user himself to handle this onerous chore. In this way, the user would be able to replay the intermediate results at some later time without the time-consuming need to rerun the entire simulation.

### *2.4 Changes*

Since many neural models involve the study of a number of related cases in which one or more parameters relating to the model are changed, it is important to give the user the ability to change parameters selectively in an easy manner. Thus, a fourth consideration in the design of CORETEX is to incorporate some way of modifying selected parameters by mentioning their name and new value during a simulation run, without having to rewrite and rerun the entire model.

### *2.5 Output*

As mentioned earlier, most neural models generate a very large volume of data which must then be transmitted to the researcher for analysis. Traditionally, this data transfer has been accomplished via some sort of mechanical device, such as a teletype or line printer. The user is then faced with the not-so-enviable task of reading this large volume of numerical output before he can understand the results of his simulation. The CORETEX language will include simple and efficient graphic representation methods to speed up this data transfer problem and make the results much easier to assimilate. This graphic representation is thus a central feature to be incorporated in our proposed neural simulation language. (See Figures 4, 5, 6, and 7.)

*2.6 Debugging*

Another desired capability of a neural simulation language is that of a built-in debugging tool. Since most nontrivial programming usually involves a substantial amount of debugging, it makes sense to include a powerful debugging aid directly in the proposed language.

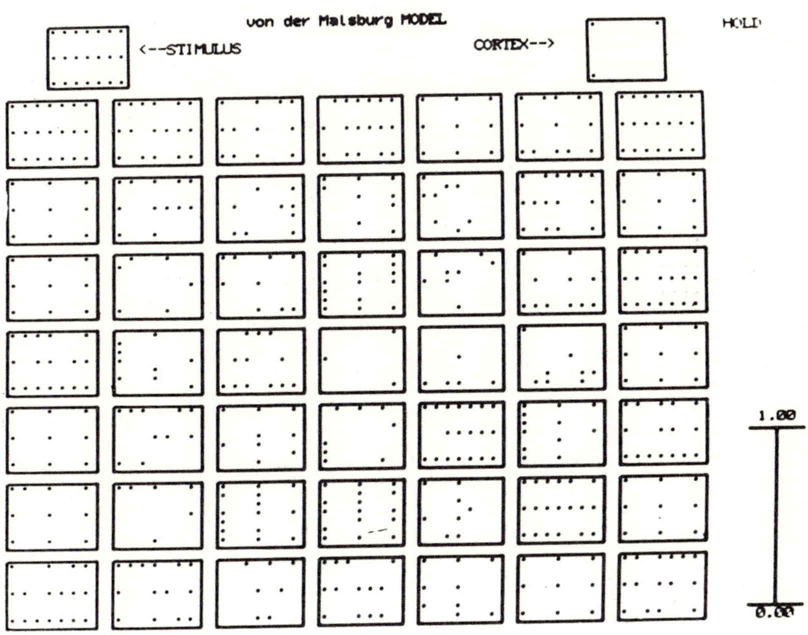


*Figure 4* A computer graphics representation of the internal state of a version of the von der Malsburg (1973) model. See Figure 8 for program listing.

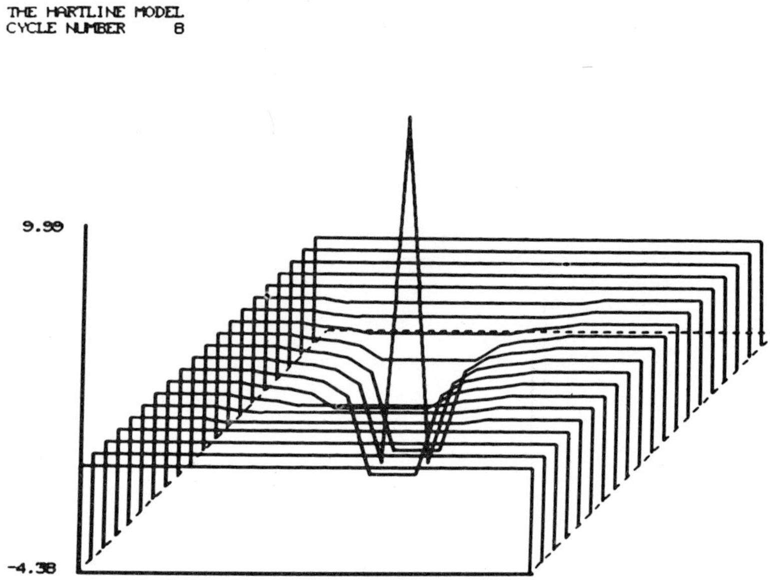


*Figure 5* A three-dimensional computer graphics representation of the pattern of activity in receptors of the *Limulus* eye. (After the experimental results of Hartline and Ratliff, 1957.)

*Figure 6* A computer graphics representation of the results of a simulated metacontrast experiment in the Amari, Lieblich, and Karshmer model (see Amari, Lieblich, and Karshmer, this volume).

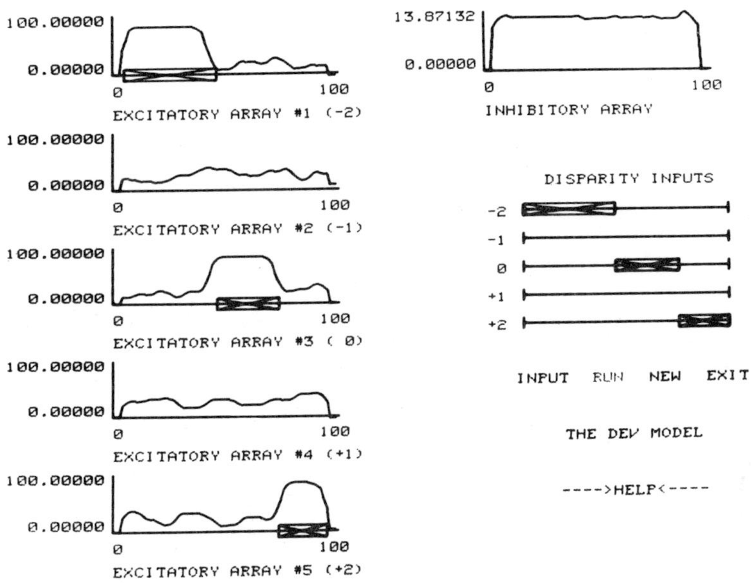


*Figure 7* A computer graphics representation of the internal state of the Dev (1974) model.

*2.7 Simplicity*

The final, but certainly not least important, consideration discussed in this section is that of simplicity of use. In fact, to become widely accepted, the CORETEX language must be easier to use and more powerful than the other alternatives available to the neural modeler. To achieve this end, we must design our language in a way that will allow the modeler to express his problem in familiar terms rather than in complicated and mystical notations (this problem is well known to any of those unfortunate people who have tried to learn LISP in an evening).

*3. IMPLEMENTATION CONSIDERATIONS*

Once the language designer has determined which features are to be included in his new language, he must finally come to grips with the problem of implementation. In fact, the implementation issue relates directly to the design stage of language development. In this section, we will discuss several of the issues which weigh heavily in the decision as to the type of implementation.

The first and possibly the most crucial issue to be faced by the language implementer is that of translation method. Translation may be defined as a mapping from the source or user-level language to the target or machine-level language. In our current discussion, we will deal with two types of translation--interpretation and compilation.

*3.1 Interpretation*

An interpreter is a program which takes program statements written in some source language and then executes the proper machine instructions to carry out the functions called for in the source statement. As soon as one statement is interpreted and executed, the interpreter goes to the next source statement and repeats the interpretation process. Indeed, if a specific statement is to be executed 1000 times in an interpreted language, it must be read and interpreted 1000 times. While this scheme may sound somewhat inefficient--which it is, it does, however, offer some very powerful advantages over other forms of translation.

One of the most powerful features available in an interpreted language is that "binding time" is delayed until the last possible moment. What this means to the user is that data types may change during program execution. In APL, for example, we could have the following four statements executed in succession with no problems:

```
[1]   A ← 'THIS IS A STRING'
[2]   A ← 5.235
[3]   A ← 15   27   53.47   199
[4]   A ← 2   2ρA.
```

In statement [1], A becomes a vector of characters upon execution because its data type is determined by its usage. In statement [2], we now change A's data type to scalar. In statement [3], A becomes a numeric vector and, finally, in statement [4], A becomes a 2 x 2 numeric matrix. After each successive statement is executed, the previous definition of A disappears. This dynamic quality of an interpreted language is, of course, not available in a language in which each data type must be bound before program execution. As a result of the dynamic nature of interpreted languages, the user gets a built-in extra feature, namely, a powerful debugging system. In an interpreted language, it is usually quite easy to stop program execution at any time, look at any variable's value, change that value, and then continue execution at the point where it was suspended.

While interpreted languages are certainly more flexible, they do suffer from one major problem--they are generally quite inefficient. This inefficiency is manifested in two crucial areas--speed of execution and memory space requirements. The loss of execution speed is due to two main factors. First, every statement must be interpreted before being executed, which is, of course, time consuming; and secondly, more than half of the time spent by the interpreter is taken up with internal housekeeping chores necessary to support the dynamically changing data types. The excess memory space required for an interpreter is due mainly to the fact that the source program, interpretive logic, and an extensive symbol table must remain resident during the interpretation process. In spite of the inefficiencies inherent in interpreted languages, they have become very popular because of their dynamic nature. Some examples of the most popular interpreted languages are APL, LISP, BASIC, and SNOBOL.

### *3.2 Compilation*

The other very popular form of translation is the compiler, which takes source or user language statements and translates them into assembly language. The assembly language code is then assembled into machine instructions. During its first pass, the compiler detects any syntactic errors which might be present. In fact, successful compilation is dependent on the absence of syntactic errors. Once compiled, the user program is translated into machine code, which can then be directly executed by the system hardware.

The compilation process buys essentially two benefits--speed and reduced memory space requirements. The compiled program runs faster simply because the source program statements do not have to be reinterpreted upon each execution as with an interpreter. Another factor which makes some compiled programs run faster is that many compilers contain logic which can actually optimize the user's source code. The memory requirements for compiled programs are generally smaller because the interpreter is not needed at run time, and there is no need for the extensive memory management logic required by most interpreters.

One of the major drawbacks to compiled languages is that they tend to be static in nature. This means that once a variable has been declared, either explicitly or implicitly, it can never change. This problem is due to the fact that the compiler must know each variable's data type before it is able to generate the appropriate assembly language code. Here we can see a good example of early binding time as opposed to the late binding time inherent in interpreters.

A second problem is that, due to this early binding time, most compilers do not load a copy of the symbol table with the actual running version of the program. Without a symbol table present, it is not only more difficult to have extensive debugging tools available to the programmer during program execution, but it is also hard to change parameters selectively during a run--a feature which we would like to incorporate in the CORETEX language.

In spite of their obvious weaknesses, compiled languages are very widely used because they run much faster, coupled with the fact that most data processing applications do not even need the dynamic features available in interpreted languages. This point is proven by the fact that most programs written in the world are for business applications, which have no need for the benefits gained through the interpretive process. Examples of compiled languages are COBOL (the most widely used computer language in world) and the old standby, FORTRAN.

Given that we would like CORETEX to be dynamic, fast running, easily debugged, and require as little memory space as possible, we are forced to search for an implementation strategy which combines the best of interpretation and compilation. This combination of translation techniques is currently used in several high-level programming languages such as SPITBOL, which is a very fast version of SNOBOL with execution speeds approximately an order of magnitude faster than that of SNOBOL.

In the combined compilation and interpretation technique, the source program is run through a compiler, which compiles as much of the program as possible into assembly language. The more dynamic parts of the source code are translated into subroutine calls to what is called an "object time" system, which carries out the memory management functions necessary. The nondynamic portions of the user program are then executed in machine code without need for reinterpretation prior to each execution.

## 4. *STRUCTURAL DESCRIPTORS*

To make CORETEX a truly powerful neural simulation language, we must provide the user with a simple method of describing the actual structure of his model--simpler, that is, than the methods available in other programming languages. Once the basic structural components common to most neural models have been identified, we must then build certain features into CORETEX which will minimize the effort needed by the researcher to describe his model. Hopefully, the CORETEX language will provide the user with the means to describe his model in terms which will be similar to a verbal description. In the discussion that follows, we list and briefly describe the most obvious of the structural descriptions to be incorporated into the CORETEX language.

*4.1 Basic Structure*

The first and most obvious item to be described by the user is the basic structure of his network. Is the model to simulate a single neuron, a one-dimensional line of cells, or a two-dimensional layer of cells? Is the layer of cells to be square, rectangular, or hexagonal as in the original von der Malsburg (1973) model?

*4.2 Connectivity*

Secondly, the user must have a simple means of describing the intercellular connectivity. For example, the modeler might like to describe a connectivity template in cases where the cellular connectivity is highly redundant, or perhaps to be able to specify the type of random connections to be employed.

*4.3 Cell Dynamics*

Next, the modeler must be able to describe the dynamics of the neurons in each layer. Are they discrete-time elements, McCulloch-Pitts (1943) neurons, or first-order "leaky integrators", or is a full-blown asynchronous, pulse-processing model desired? It would be desirable to allow the user to simply select the proper cell dynamics from built-in language primitives, but if he so desires, it should also be possible for him to specify his own cell dynamics.

*4.4 Synaptic Dynamics*

In addition to the cellular dynamics, the neural modeler must specify the "synaptic dynamics" of his model--i.e., the way in which synaptic strengths vary with time and events in the network. In the simplest case, the researcher might choose to fix the synaptic strengths, but when describing adaptive models, the modeler might require habituating connections, Hebbian synapses (Hebb, 1949), or some other type of connectivity.

*4.5 Stimuli*

Another important structural item which the neural modeler must define is the input stimuli. The CORETEX language should facilitate the specification of what types of stimuli are to be applied as input to the network and the manner in which the simulation is to be run.

### *4.6 Output*

In order to make the data generated by his model meaningful, the user must be given a simple means of describing the form of output which will best illustrate the results obtained while running the model. The decision on desired output specification should be made during the structural specification stage of modeling; therefore, the CORETEX language should contain built-in data display primitives.

### *4.7 Initial Conditions*

Finally, during the structural description phase, the user should be given a simple method of describing the initial condition of the cellular and synaptic internal states as well as the accuracy required of the integration procedure.

## *5. AN EXAMPLE*

Although CORETEX remains in the design phase and has not as yet been implemented, some flavor of the type of language that we seek to create can be given by the following example. A recent model of the development of visual cortex by von der Malsburg (1973) consists of two two-dimensional layers, one representing the retina, the other visual cortex. Every retinal receptor projects to each cortical cell. The cortical cells have excitatory projections to their nearest neighbors in cortex and inhibitory projections to cortical cells that are located farther away. Von der Malsburg showed that when the synapses from retina to cortex are assumed to be Hebbian and when the model is shown, in turn, straight lines of each of a number of orientations, each cortical cell will eventually develop a preferred orientation. In fact, cells that are close together in cortex will have similar orientation preferences. A possible FORTRAN encoding of this model can be seen in Figure 8.

```
RK1:VDM.FOR

C
C---->CREATE WORKING ARRAYS
C
      DIMENSION CORT(7,7),S(7,7,7,7),ISTIM(7,7,9),STIMP(7,7)
      DIMENSION DISTAR(7,7,7,7),LP(7)
      INTEGER DFILE(4000)                  !**ESTABLISH DISPLAY FILE
      DATA NO/'N'/
      DATA ISTIM/7*1,14*0,7*1,14*0,7*1,1,0,0,1,0,0,1,
      1        1,0,0,1,0,0,1,
      2        1,0,0,1,0,0,1,
      3        1,0,0,1,0,0,1,
      4        1,0,0,1,0,0,1,
      5        1,0,0,1,0,0,1,
      6        1,0,0,1,0,0,1,343*0/
      PLUS(A)=(A+ABS(A))/2.0                        !**THE PLUS FUNCTION
C
C---->INITIALIZE DISPLAY FILE
C
      CALL RZERO(STIMP,49)

      CALL INITDF(DFILE,4000)
      CALL APNT(300,1013,-1,0,0,DFILE)          !**PUT TITLE ON SCREEN
      CALL TEXT(4,0,0,1,'VON DER )',DFILE)
      CALL TEXT(4,0,0,0,'M)',DFILE)
      CALL TEXT(4,0,0,1,'ALSBURG )',DFILE)
      CALL TEXT(4,0,0,0,'MODEL)',DFILE)
      JEND=NEXT(DFILE)-1
      CALL LPEN(LP)                             !**SET UP LIGHT PEN
      LP(1)=0                                   !**ENABLE LIGHT PEN
C
C---->GET STIMULI
C
      IF (LINK().EQ.1)CALL SCROL(10,0,1000)     !**GET RID OF SCROLLER
      NSTIM=2                                   !**FOR STANDARD CASE
      IDIM=7
      WRITE(7,25)
25    FORMAT(' DO YOU WANT A NEW STIMULUS SET ?',$)
      READ(5,50)NEWOLD
50    FORMAT(A1)
      IF(NEWOLD.EQ.NO)GO TO 400
      WRITE(7,75)
75    FORMAT(' SIZE OF RETINA AND CORTEX ? (MAX. IS 7) ',$)
      READ(5,200)IDIM
      WRITE(7,100)
100   FORMAT(' HOW MANY STIMULI IN SET ?(MAX. IS 9) ',$)
      READ(5,200)NSTIM
200   FORMAT(I1)
      DO 300 K=1,NSTIM
      WRITE(7,250)K
250   FORMAT(//' STIMULUS NUMBER ',I1/)
      DO 300 I=1,IDIM
300   READ(5,350)(ISTIM(I,J,K),J=1,IDIM)
350   FORMAT(7I1)
C
C---->SET INITIAL SYNAPTIC WEIGHTS AND DISTANCE FACTORS
C
400   CALL START(DFILE)
      IF(LINK().EQ.1)CALL SCROL(1,0,1)
      IX1=0
      IX2=0
      DO 450 I=1,IDIM
      DO 450 J=1,IDIM
```

*Figure 8* A FORTRAN encoding of a version of the von der Malsburg (1973) model.

```
         DO 450 K=1,IDIM
         DO 450 L=1,IDIM
         DISTAR(I,J,K,L)=SQRT(FLOAT((K-I)**2+(L-J)**2))
450      S(I,J,K,L)=RAN(IX1,IX2)
C
C---->SET OTHER CONSTANTS
C
         TAUS=1.0
         TAUC=1.0
         NCYCLE=10
         DELT=0.1
         DTONTS=DELT/TAUS
         DTONTC=DELT/TAUC
C
C---->MAIN LOOP
C
500      DO 1500 IWHICH=1,NSTIM
         DO 550 K=1,IDIM
         DO 550 L=1,IDIM
         STIMP(K,L)=FLOAT(ISTIM(K,L,IWHICH))
550      CORT(K,L)=0.0
C
C---->STIMULUS LOOP
C
         DO 1500 ICYCLE=1,NCYCLE
C
C---->CORTICAL CELL LOOP
C
         DO 600 K=1,IDIM
         DO 600 L=1,IDIM
         CORTP=CORT(K,L)
         DRIVE=0.0
         COMP=0.0
C
C---->RETINAL LOCUS LOOP
C
         DO 575 I=1,IDIM
         DO 575 J=1,IDIM
         STIMPP=STIMP(I,J)
         SP=S(I,J,K,L)
         DRIVE=DRIVE+STIMPP*SP
         DIST=DISTAR(I,J,K,L)
         IF(DIST.EQ.0.0)GO TO 575
         COMP=COMP+PLUS(CORT(I,J))/DIST
575      S(I,J,K,L)=SP+STIMPP*PLUS(CORTP)*DTONTS
600      CORT(K,L)=CORTP+(DRIVE-CORTP-COMP)*DTONTC
C
C---->NORMALIZE SYNAPTIC WEIGHTS
C
         DIMSQ=FLOAT(IDIM*IDIM)/2.0
         DO 750 K=1,IDIM
         DO 750 L=1,IDIM
         STOT=0.0
         DO 700 I=1,IDIM
         DO 700 J=1,IDIM
700      STOT=STOT+S(I,J,K,L)
         DO 750 I=1,IDIM
         DO 750 J=1,IDIM
750      S(I,J,K,L)=S(I,J,K,L)*DIMSQ/STOT
         CALL DRETN(DFILE,JEND)
1500     CALL TMAP(S,7,7,7,7,IDIM,IDIM,IDIM,IDIM,3,0,DFILE,
         1    4000,0.50,CORT,STIMP,7)                     **PUT UP DISPLAY
         GO TO 500
         STOP
         END
```

*Figure 8* (continued)

In CORETEX, a much more readable and simple expression is possible. One would require only the following statements:

```
MODEL VDM
REAL RETINA (5, 5), CORTEX (5, 5) --- (no extra cells)
RETINA = INPUT --- (no dynamics)
CORTEX = LI (1.0, RETINA, CORTEX)
RETINA TO CORTEX:  RANDOM (UNIFORM, 0.04),
          NORMALIZED HEBBIAN (10.0)
CORTEX TO CORTEX:  GAUSSIAN (-0.5, 1)
STIMULUS A = (5*1, 5*0, 5*1, 5*0, 5*1)
STIMULUS B = (5* (1, 0, 1, 0, 1))
DISPLAY (THRESHOLDED (0, 1)):  RETINA, CORTEX,
          (RETINA TO CORTEX)
/RESET/
SET CORTEX TO 0.0
SET (RETINA TO CORTEX) TO RANDOM (UNIFORM, 0.04)
$
END MODEL
```

## 6. *CONCLUSIONS*

A language of the sort proposed here would have a number of benefits. It would, of course, ease the task of programming for those interested in simulating neural nets. This might have the desirable side effect that experimentalists who normally shy away from programming a computer would much more easily be induced to undertake simulation studies related to their experimental research. Some effort will be made to couch the language in sufficient generality so that other parallel processes, in addition to brain models, could be simulated on the system. Thus, we hope that workers in artificial intelligence might also find CORETEX of some use.

*REFERENCES*

Blake, K., and Gordon, G. 1964. Systems simulation with digital computers. *IBM Systems Journal III, 1:* 14-20.

Dahl, O.-J., Dijkstra, E. W., and Hoare, C. A. R. 1972. *Structured Programming*. London: Academic Press.

Dev, P. 1974. Segmentation processes in visual perception: A cooperative neural model. Department of Computer and Information Science Technical Report 74C-5. University of Massachusetts at Amherst.

Efron, R., and Gordon, G. 1964. A general purpose digital simulator and examples of its application. I. Description of the simulator. *IBM Systems Journal III, 1:* 22-34.

Hartline, H. K., and Ratliff, F. 1957. Inhibitory interaction of receptor units in the eye of *Limulus*. *Journal of General Physiology 40:* 357-376.

Hebb, D. 1949. *The Organization of Behavior*. New York: John Wiley and Sons.

McCulloch, W. S., and Pitts, W. 1943. A logical calculus of ideas immanent in nervous activity. *Bulletin of Mathematical Biophysics 5:* 115-133.

von der Malsburg, C. 1973. Self-organization of orientation sensitive cells in the striate cortex. *Kybernetik 14:* 85-100.

# Motivational Learning of Spatial Behavior

MICHAEL A. ARBIB
*Department of Computer and Information Science*
*and*
*Center for Systems Neuroscience*
*University of Massachusetts at Amherst*

ISRAEL LIEBLICH*
*Center for Systems Neuroscience*
*University of Massachusetts at Amherst*

ABSTRACT *A model for motivational learning of spatial behavior is presented. It postulates a spatial representation in the form of a directed graph with nodes containing drive-related information and edges containing sensorimotor features. This representation is called the World Graph (WG). The model specifies the general drive dynamics, the ways in which WG is updated, and the ways in which the organism moves about in the world. Two hypotheses on the interaction of choice of nodes and drive levels are discussed. Two hypotheses on the way in which weights are assigned to the different nodes in the WG are developed. The model is constrained by experiments by Brown (1942, 1948), Gough (1962), Hull (1933), Leeper (1935), Miller (1959), and Strain (1953).*

## 1. *INTRODUCTION*

The tendency toward conceptual divergency of human memory studies and animal learning and conditioning research has been noted by Estes (1973). As a first stage in reducing the gap between these two bodies of knowledge, he envisaged conditioning as the

---

*1975-76 Fellow, on leave from the Department of Psychology, The Hebrew University of Jerusalem, Jerusalem, Israel.

establishment of a hierarchical structure in memory. This structure incorporates representation of stimuli in their contexts and associations between these representations.

While this suggestion has the flavor of approaching the language of human memory theorists and thus may serve as a first step towards a link with animal learning theorists, it still does not make any concrete suggestions as to the ways in which an organism couples its memory structure to its specific action routines such that it might operate in its spatial environment in a meaningful fashion. The lack of specific theories in this area should not come as a surprise. Most of the experimentalists working with the classical conditioning and the bar pressing/operant conditioning paradigm are not interested in moving the animal in space. Thus, as a first approximation, we are left with the fortunate situation of being quite free in suggesting assumptions and being constrained only by a relatively small number of experiments!

By choosing to specifically make our organism move in space, we force ourselves to address many of the critical elements in a behavior model.

## 2. *THE WORLD GRAPH*

We wish to analyze how an animal builds up a model of the world around it in such a way that, when under the influence of a particular drive, it can find its way to a location where it has learned that the drive will be reduced.

What form is the model to take? In particular, how is a place known to the animal to be internally represented? It seemed to us to prejudge the issue to imagine that the animal's primitive representation of such a place should include the Euclidean coordinates of the place. We did not wish to preclude the *development* of a Euclidean frame through experience, but we wished to show that a far less structural representation would suffice to explain much spatial behavior, at least in the rat.

After some experimentation, then, we have posited that the

animal's spatial representation takes the form of a directed graph.

The *world graph* has a set of nodes, with each node x corresponding to a *recognizable place* in the animal's world. It is an important feature of our model that a single place in the world may be represented by more than one node in the graph--if, for example, the animal comes upon a place in the maze for the second time but does not "recognize" that he has been there before.

There is an edge from node x to node x' in the graph for each distinct path the animal has traversed from the place he recognizes as x to the place he recognizes as x' which does not pass through another recognizable place.

In the model that we present here, the world graph has sensorimotor features associated with the corresponding path (e.g., turn left towards the +) appended to each edge--while the nodes contain information about expectation of drive reduction R. This is because we posit that any place where notable reinforcement occurs, for example, will be represented by its own node, rather than simply remaining somewhere in limbo between two recognizable places.

It should be noted that we do not equate drive reduction with reinforcement--saccharin may reinforce as much as sugar. But we posit that both affect the animal's expectation of reduction of hunger. We will distinguish an *actual* drive-reduction $a_1$ (e.g., eating food) from a drive-related reinforcement $a_2$ (e.g., taking saccharin).

Before describing the information stored with a node of the graph, we must say something about drives and their dynamics.

## 3. *DRIVE DYNAMICS*

We posit a set $d_1, d_2, \ldots, d_k$ of discrete *drives* which control the animal's behavior. At time t, each drive d in $(d_1, \ldots, d_k)$ has a value d(t). Typical d's are thirst, hunger, sex, and fear. *Appetitive* drives like thirst and hunger are represented by positive numerical values: $0 \leq d(t) \leq d_{max}$; while *avoidance* drives like fear are represented by negative numerical values: $d_{max} \leq d(t) \leq 0$. We say a drive *increases* if it changes towards $d_{max}$, and is *reduced* if it changes toward 0. Thus, these terms properly apply to the drive *magnitude*, $|d(t)|$.

We must now indicate how the level of any drive changes over time. We shall adopt a discrete time scale, $t = 0,1,2...$ . For the time being, we may think of the time unit as being the time required to move from one recognizable place (i.e., a place represented by a node of the graph) to another. It is only in Section 7 that we will need a finer time grain, corresponding to an actual step or two of the animal's feet.

The idea is that each drive spontaneously increases with time. A reduction takes place in the presence of some substrate --such as food in the case of hunger--for *drive reduction* (our apologies for the circularity). An additional increase occurs if an *incentive*--such as the aroma of food in the case of hunger--is present. We thus posit the following quantities for each drive d:

$\alpha_d$ = the spontaneous rate of increase of $d(t)$,

$a_1(d,x,t)$ = the reduction factor upon drive d of the place represented by node x at time t,

$a_2(d,x,t)$ = the reinforcement related to drive d at the place represented by node x at time t,

$I(d,x,t)$ = the incentive factor for drive d of the place represented by node x at time t.

Each of these quantities is positive for an appetitive drive d; negative for an avoidance drive.

Let us use P(x) as shorthand for "the place represented by node x." We reiterate that our model allows $x \neq x'$ even if $P(x) = P(x')$. It is reasonable to postulate that $P(x) = P(x') \Rightarrow a_1(d,x,t) = a_1(d,x',t)$ and $I(d,x,t) = I(d,x',t)$ for all drives d and times t, but we shall not make explicit use of this postulate in the paper.

Note that each $\alpha_d$ is to be regarded as a parameter intrinsic to the animal, while--for an animal in place P(x) at time t--the quantities $a_i(d,x,t)$ and $I(d,x,t)$ for each drive d are to be regarded as representing the current environmental stimulation received by the animal. Other stimuli provide the cues which enable the animal to recognize the place it is in as being represented by node x, but this place recognition problem will receive only brief mention in what follows.

We now postulate the following form for the drive dynamics: If an animal is at the place it recognizes as being node x at time t, and the value of drive d at that time is d(t), then the value of d for the animal at time t + 1 will be

$$\begin{aligned} d(t+1) = d(t) &+ \alpha_d |d_{max} - d(t)| && \text{spontaneous change,} \\ &- a_1(d,x,t)|d(t)| && \text{reduction,} \\ &+ I(d,x,t)|d_{max} - d(t)| && \text{incentive.} \end{aligned}$$

We do not insist on the exact form of this equation, but stress two useful properties: (i) These dynamics maintain drive levels within the specified range between 0 and $d_{max}$ so long as neither $\alpha_d + I$ nor $a_1$ exceed 1 in absolute value; and (ii) The dependence on drive reduction, for example, has the right qualitative shape--if we think of $a_1(d,x,t)$ as the amount of food eaten, it makes sense to think of the actual amount of food eaten increasing with both $a_1(d,x,t)$ and d(t). More realistic formulas are possible, but are not necessary for the somewhat qualitative findings presented in this paper.

In this particular model, there is not complete forgetting in the sense of all traces of a node being removed from the world graph. However, if the organism constantly meets nodes with higher R values than those of a given x, then--by the rules to be given in Section 5--the animal's behavior becomes less and less likely to show overt knowledge of x.

## 4. UPDATING THE WORLD GRAPH

We have now postulated that the state of the animal at any time comprises *at least* the vector $[d_1(t),\ldots,d_k(t)]$ of its drive levels and its current world graph WG(t). We have postulated how the drives change over time in response to environmental stimulation. However, we have yet to describe what information is appended to the nodes of the world graph, or how the world graph changes over time.

We were careful to distinguish each node x of the graph from the place P(x) which it represents. In the same fashion, we must distinguish the *observable* reduction $a_1(d,x,t)$ of drive d which the animal will actually experience if it is at node x at time t from the *expected* reduction R(d,x,t): R(d,x,t) is the animal's internal representation, based on experience prior to t, of the extent to which substrates at x serve to reduce drive d, or at least provide reinforcement related to drive d.

We then postulate that, attached to each node of x of WG(t), is the vector $[R(d_1,x,t),\ldots,R(d_k,x,t)]$ of the animal's current expectations at time t about the drive-related properties of P(x).

Note that whereas P(x) = P(x') implies $a_1(d,x,t) = a_1(d,x',t)$ for each d and t, it is *by no means* the case that P(x) = P(x') implies R(d,x,t) = R(d,x',t) for a given d and t. In fact, x may correspond to "door A when there is a cross on it," while x' may correspond to "door A when there is a circle on it," and so R(d,x,t) may be justifiably large, while R(d,x',t) may be justifiably small.

This completes our specification of what the world graph *is* at any time. We must now specify how it *changes*. The changes are of two kinds: changing the R values labelling the nodes and actual structural changes in the graph.

We postulate that R(d,x) does not change if the animal is not at node x:

$$R(d,x,t+1) = R(d,x,t) \text{ if the animal is } \textit{not} \text{ at node x at time t.}$$

However, if the animal is at node x, then the expected drive reduction factor R(d,x,t) will be changed toward the actual drive reduction factor, with the change being greater the more intense the value of d(t):

$$R(d,x,t+1) = \frac{d(t)}{d_{max}} a_2(d,x,t) + \left(1 - \frac{d(t)}{d_{max}}\right) R(d,x,t)$$

if the animal *is* at node x at time t.

Note that $a_2(d,x,t)$ may include both actual drive reduction as well as taking into account a short-term component (e.g., saccharin, electrical brain stimulation) in the computation of R(d,x,t). Thus, we get "one-shot learning" if d(t) is close to $d_{max}$, so that $1 - \frac{d(t)}{d_{max}}$ is close to 0, but very little change in R(d,x) if d(t) itself is close to 0.

There are two types of *structural* change: adding a node and merging nodes. If the animal at time t decides to take a path that it has not traversed before, then--in addition to the updating of R(d,x,t) to R(d,x,t + 1) just described--a *new* node x' will be added to the world graph to represent the first recognizable place along this path, together with an edge from x to x'. [Two comments: (i) In Section 6 we shall explain how a novel path comes to be traversed. (ii) A recognizable place is one with distinctive features that may make it memorable, but need not be one which has been seen before. Thus, a junction in a maze is more likely to be represented by a node in the world graph than is a place along a featureless corridor.] We posit that each drive d has a neutral reduction value $a_0(d)$--and we assign to x' the expected drive reduction factors $R(d_i,x',t+1) = a_0(d_i)$, for $1 \le i \le k$. These neutral values will then be updated when the above algorithm is applied to compute

$$R(d,x',t+2) = \frac{d(t)}{d_{max}} a_2(d,x',t+1) + \left(1 - \frac{d(t)}{d_{max}}\right) a_0(d).$$

Two nodes are merged when the animal decides that x and x' represent the same *situation*. (Note: The word "situation" is carefully chosen rather than "position"--as the following discussion shows.) It is beyond the scope of this paper to discuss state merging in detail, but the following example gives some idea of the flexibility of the notion.

Suppose that a rat has been trained in a T-maze, where the door at each end of the T-bar can bear a + or a 0, and where the + is consistently rewarded with food. Because we have not required that each location in Euclidean space be represented by a single node of the world graph, it is quite plausible that the node-creation procedure will add to $x_0$--the node representing the start box--not two but four nodes:

(L,+), (L,0), (R,+), (R,0),

where (L,+) is "the door to the left with a + on," etc. As a result of its training, the animal may well have R(hunger,x,t) close to 1 for (L,+) and (R,+), and close to zero for (L,0) and (R,0). Because of their commonalities in both sensory features and R(hunger,x) values, (L,+) may merge with (R,+), and (L,0) may merge with (R,0). In short, the world graph representation is rich enough to allow generalization over dimensions other than the spatial.

Deutsch (1960) has mechanically implemented a similar scheme for a prespecified set of nodes, but did not have the concept of a modifiable world graph.

## 5. *MOVING ABOUT THE WORLD*

We have now completely specified the *internal state of the animal* at time t. It comprises three components:

| | |
|---|---|
| The drive vector | $[d_1(t),\ldots,d_k(t)]$; |
| The world graph | WG(t); |
| The current node | x(t) of WG(t) which is the animal's internal representation of its position. |

We have said that the *environmental stimuli* to the animal when at the place represented by node x are represented by the drive-related factors $a_j(d_i,x,t)$ and the incentive factors $I(d_i,x,t)$, $1 \le i \le k$, as well as environmental cues which let the animal determine whether it is in a situation represented by a node of WG(t), or whether a new node must be added to form WG(t + 1).

Moreover, we have given rules for *updating the internal states*--specifying the drive dynamics in Section 3, and the updating of the world graph in Section 2. These specifications rest on no empirical data but--we have suggested--are quite robust.

What we have not yet specified is the *output* of our system (Figure 1)--the process whereby the animal decides *where to move* next. A fuller model would also describe how the animal's consummatory behavior affects $a_1(d,x,t)$, but for now we assume these values are completely under the control of the experimenter.

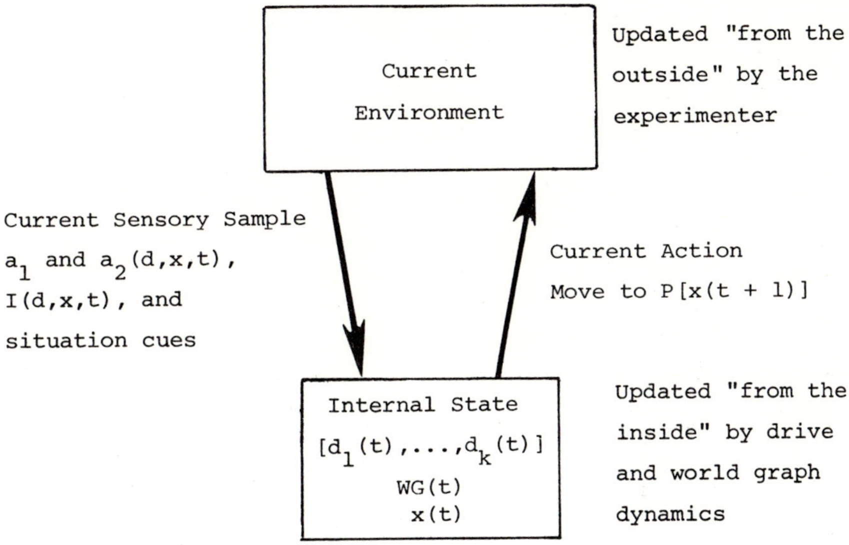


*Figure 1*

Whereas the previous specifications have been based on plausibility arguments alone, we shall use experimental data to *constrain* the form of the *next-node function* which determines where the animal will move.

We wish to distinguish between two hypotheses about the relationship between drive levels and the choice of the next node:

*The dominant drive hypothesis:* The animal goes to a node most likely to reduce its dominant drive--that is, a $d_i(t)$ such that $d_i(t) \geq d_j(t)$ for $1 \leq j \leq k$.

*The competing nodes hypothesis:* The animal goes to a node with the greatest expected reduction of all drives.

We shall present experimental evidence due to the work of Brown (1942, 1948) and Miller (1959) which suggests the latter hypothesis. We next distinguish two hypotheses about chaining:

*The local hypothesis:* The weights assigned a node x in the node-selection process depend only on the R(d,x,t) for that x.

*The chaining hypothesis:* The weights assigned a node x in the node-selection process depend not only on the R(d,x,t) for that x, but also on the R(d,x',t) for nodes x' reachable on paths through x.

We shall present experimental evidence which Gough (1962) offered in support of the local hypothesis, and evidence due to Strain (1953) which requires the chaining hypothesis. However, we shall see that the Gough data are consistent with the chaining hypothesis once it is realized that the same place may be represented by more than one node in the animal's world graph.

We introduce the notation $\hat{d}(t)$ for the set of *operative drives* at time t.

$$\hat{d}(t) = \begin{cases} \text{a } d_i \text{ such that } d_i(t) = \max[d_j(t) \mid 1 \leq j \leq k] \\ \quad \text{if the dominant drive hypothesis holds,} \\ (d_1, \ldots, d_k) \text{ if the competing nodes hypothesis} \\ \quad \text{holds.} \end{cases}$$

We next introduce the *augmented expected drive reduction factor* $\hat{R}(d,x,t)$:

$$\hat{R}(d,x,t) = \begin{cases} R(d,x,t) \text{ if the local hypothesis holds,} \\ \text{some combination of } R(d,x,t) \text{ with } R(d,x',t) \\ \text{for (some of) the nodes } x' \text{ reachable from} \\ x \text{ if the chaining hypothesis holds.} \end{cases}$$

In fact, if the chaining hypothesis holds, then the evaluation of a successor node x when made at a node $x_0$ may depend only on nodes x' which are beyond x when moving from $x_0$.

With this terminology, we are ready to state our hypothesis about how the animal chooses its next move:

Let a rat be at the place it recognizes as represented by node x. For each node x' immediately adjacent to x compute

$$b(x',t) = \Sigma \{\hat{R}(d,x',t)d(t) \mid d \in \hat{d}(t)\}$$

[which reduces to just $R(d_i,x',t)d_i(t)$ if both the local hypothesis and the dominant drive hypothesis hold, and $d_i$ is the dominant drive].

There is a confidence level ε such that the rat will always go to a node x' if its b-value exceeds that of all others by ε. However, if several nodes fall within the top ε of b-values, we posit that the choice is made at random:

$$\text{Let } b = \max b(x',t).$$

$$\text{Let } \tilde{x}(t) = \{x' \mid x' \text{ is adjacent to the current node } x, \text{ and } b(x',t) \geq b - \varepsilon\}$$

Then the next node, $x(t + 1)$, is selected at random from $\tilde{x}(t)$.

As usual, we do not insist on the specific form of this prescription, but we do stress two qualitative features: (i) By multiplying $R(d,x',t)$ by $d(t)$ in the definition of $b(d,x,t)$, we ensure that if a drive is very much stronger than the others, it will dominate the node-selection process even if the dominant drive hypothesis does *not* hold. This feature would ensure that the experimental results related to state-dependent learning (e.g., Brown, Feldman, and Moore, 1968; Hull, 1933; Leeper, 1935) would be reproduced by this model. (ii) By introducing $\varepsilon$, we ensure that the animal will not repeatedly choose the same node unless it has been exceptionally well rewarded there.

[We wish to suggest a feature not necessary for the present exposition which might have potential for future predictions at a more neural level--the Drive Interaction Matrix (DIM). We define $DIM_{d,d'}$ such that:

$$DIM_{d,d} = 1 \text{ for all } d,$$

$$DIM_{d,d'} \begin{cases} > 0 \text{ if } d'\text{-drive adds to } d\text{-drive,} \\ < 0 \text{ if } d'\text{-drive inhibits } d\text{-drive.} \end{cases}$$

The effective strength of a drive d at time t is then:

$$effd(t) = \sum_{d'} DIM_{d,d'} \cdot d'(t).$$

The dominant drive is the one with max effd.]

An important feature of this model is that it yields *exploration* and *latent learning without the introduction of an explicit exploratory drive*. The animal will "explore" unless there is a node x' whose b(x',t) is markedly higher than that of all adjacent nodes, and this is unlikely to happen unless one drive is currently much higher than the others.

We need one final definition: What is meant by a node x' *immediately adjacent* to a node x? Certainly, if there is an edge from x to x' in WG(t), then x' is immediately adjacent to x. However, in Section 4 we specified that new nodes be added to the graph. The idea is that we also consider as adjacent to x, world-situations $\hat{x}$ not yet represented in the graph, but which can be reached from node x--such as "the door with a + on it, to the left of P(x)." Any of these situations may have their $b(\hat{x},t)$ computed, using the neutral values $R(d,\hat{x},t) = a_0(d)$. If, as a result, $\hat{x}$ becomes $x(t + 1)$, then a new node $\hat{x}$ will be added to the world graph, and the edge from x to $\hat{x}$ will be tagged with the appropriate defining features. If, on reaching $P(\hat{x})$, the animal recognizes a situation already represented in the graph by some node $x_1$, then $\hat{x}$ will be merged with $x_1$, which will now have an extra edge leading to it from x. If the animal does not move into a new situation at time t, there will be no node added to WG(t) in forming WG(t + 1).

We now turn to our arguments in favor of the chaining hypothesis and the competing nodes hypothesis.

## 6. *THE CHAINING HYPOTHESIS*

We first present an experiment due to Gough (1962) which was offered as evidence against the chaining hypothesis, and then show that it is in fact consistent with the chaining hypothesis. We then present an experiment due to Strain (1953) which does not seem interpretable without some form of the chaining hypothesis.

A rat was trained in a maze like that indicated in Figure 2. During training, barriers were placed at B and B' so that the rat could not get from A to C. The rat learned that it could go from A to B or B' to get food, and from B to C to get food. Moreover, the drive reduction at C was greater than at B or B'.

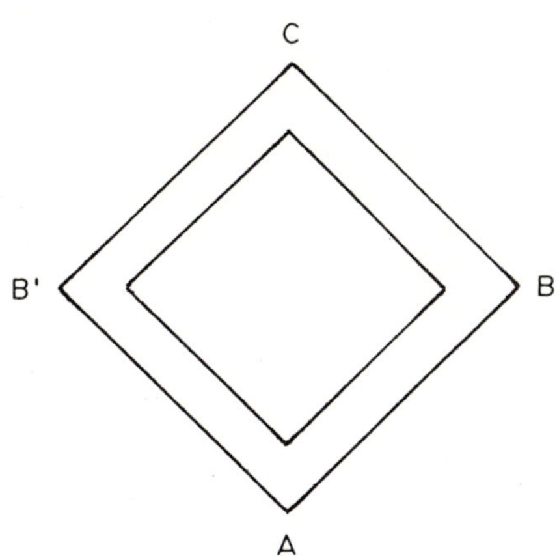


*Figure 2*

After training, the barriers were removed. To use our terminology, Gough argued that if the chaining hypothesis were correct, then the effect of the high value of R(hunger,C,t) would cause $\hat{R}$(hunger,B,t) to well exceed R(hunger,B',t), and so the rat should choose B over B'. In fact, in the first trial after barrier removal, rats were just as likely to choose B' as B. Gough took this as evidence of the rat's inability to chain segments it had learned separately.

We offer an alternative hypothesis. We suggest that after training the world graph of the rat had a form like that shown in Figure 3. The important point is that B is represented by two distinct nodes--one for each side of the barrier--and despite the fact that Gough took care to provide distinctive features at each junction of the maze, we see no reason why the rat should have merged $x_{BA}$ and $x_{BC}$ on the basis of its training experience.

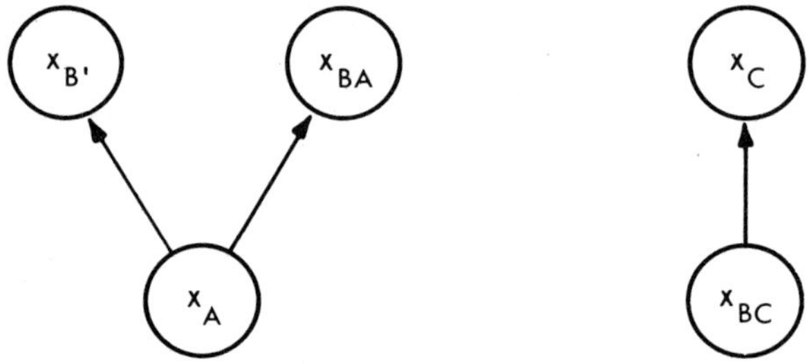


*Figure 3*

Once the rat has gone from A to B with the barriers down, Gough finds that it does indeed proceed to C (food having been removed from B and B' with the barriers), but that if it goes from A to B', it does not go to C. We would explain all the data as consistent with the chaining hypothesis--with the added proviso that, once the barriers are down, the animal can, once it has reached B, recognize $x_{BA}$ and $x_{BC}$ as representing the same situation, and so can merge $x_{BA}$ and $x_{BC}$ into a single $x_B$.

(Gough's argument was more subtle than we present here--he was not so much arguing against the chaining hypothesis per se as he was arguing against a form of it implicit in Hull's theory. The model presented here might well be acceptable to him as an attempt to move beyond too strict a Hullian formulation.)

We now turn to the work of Strain (1953). He allowed a rat to explore a linear array of "stations" such as that shown in Figure 4. According to our model, exploration and latent learning would lead to a world graph essentially isomorphic to the array of Figure 4, with neutral values for all drives except fear, which would have a relatively reduced value for each node.

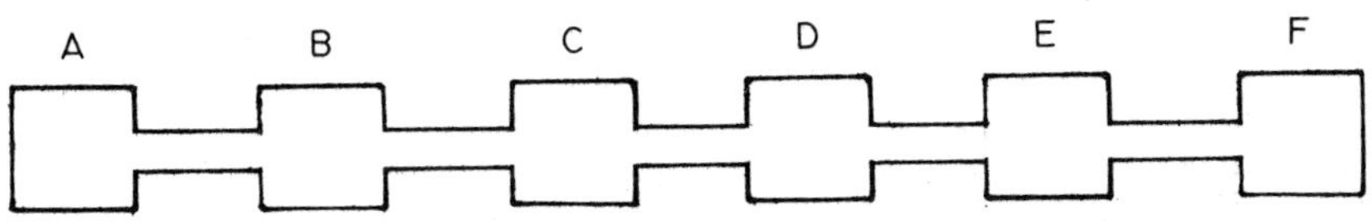


*Figure 4*

After exploration, a massive shock is administered to the rat at Station F. This causes R(fear,F) to become large and negative. It also makes the fear drive, $d_{fear}(t)$, large and negative. Thus, fear will dominate node choices whether the dominant drive hypothesis or the competing node hypothesis is correct. We thus ignore all drives other than fear in the rest of this section. Now R(fear,x) is negative for x = F and neutral for x $\neq$ F. Thus, the rat at node E will choose node D over node F. But what of the rat at B, C, or D? Strain found a consistent perference for a move to the node further from F. This supports the chaining hypothesis since the local hypothesis gives no grounds for preference in these cases.

If we accept the chaining hypothesis, we must still provide a formula for $\hat{R}(d,x,t)$ in terms of nodes beyond x. The form is unclear from the experiment, and so we offer a tentative formula:

$$\hat{R}(d,x,t) = \sum_{x' \varepsilon \Delta_x} \frac{R(d,x',t)}{\ell_d(x,x') + 1},$$

where $\Delta_x$ is the set of (some of) the nodes x' reachable beyond x (and including x) from the current state, while $\ell_d(x,x')$ is the subjective distance "how far x' seems to be from x when the dominant drive is d." The emphasis on a distance which is subjective and drive-dependent--rather than the simple use of Euclidean distance--will be justified in the next section.

Let us briefly note the implication of this formula for Strain's set up. If we posit that $\ell_d(x,x')$ takes the simple form of "number of steps separating x from x'," and if we take

$$R(\text{fear},x) = \begin{cases} -\gamma & \text{for } x = F, \\ 0 & \text{for } x \neq F, \end{cases}$$

we deduce that

$$R(\text{fear},x) = \begin{cases} -\gamma/6 & \text{for A}, \\ -\gamma/5 & \text{for B}, \\ -\gamma/4 & \text{for C}, \\ -\gamma/3 & \text{for D}, \\ -\gamma/2 & \text{for E}, \\ -\gamma & \text{for F}. \end{cases}$$

If A is to be chosen over C consistently from node B, we must have

$$(-\gamma/6) - (-\gamma/4) > \varepsilon$$

which reduces to

$$\gamma > 12\varepsilon.$$

Returning to our formula in Section 4 for updating R, we see that if we assume R(fear,F) was close to 0 prior to the shock, and if the fear drive, $d_{fear}(t)$, was relatively low prior to the shock--say $d_{fear}(t)/d_{fear,max} = 1/5$--we have

$$\gamma \cong \frac{1}{5} a_1(\text{fear},x,t),$$

so that the shock, $a_1(\text{fear},x,t)$, must satisfy

$$a_1(\text{fear},x,t) = 5\gamma > 60\varepsilon$$

if the preference effect is to show up at B. This explains why Strain had to apply large shocks to reliably obtain the effect.

## 7. *THE COMPETING NODES HYPOTHESIS*

Miller (1959) studied the behavior of a rat on a runway from A to B, with the rat encountering food and shock at B on different occasions. He found monotonically decreasing gradients of approach and avoidance with a crossover point at which vacillation takes place.

We may interpret these results in a fashion consistent with our earlier restrictions only if we adopt the competing nodes hypothesis. Raising the hunger drive considerably only moves the point of vacillation closer to B--it does not yield a state in which hunger dominates fear to the exclusion of avoidance behavior.

Let us work, then, with both the chaining hypothesis and with the competing nodes hypothesis. To handle Miller's data we must, however, abandon our coarse time-graining in which successive moments of time correspond to the animal's arrival at recognizable places. Instead, we must turn to a finer time-grain in which the unit interval is the time required for the animal to traverse some small distance $\delta y$. We then compute the animal's next move at each time t--but now we cannot simply refer to a node x to give the animal's current situation, but must instead characterize its situation by the triple $(x,y,x')$--meaning that it is enroute from $P(x)$ to $P(x')$, with distance y to go until it reaches $x'$.

To explain Miller's results, consider the situation in which $R(\text{hunger},A,t) = R(\text{fear},A,t) = 0$, whereas $R(\text{hunger},B,t) = 1$ and $R(\text{fear},B,t) = -1$. When the actual distance of the animal from B is y, let

$$f(y) = \ell_{fear}[(x,y,x'),x'],$$

which is "how far y seems to be from x' when the dominant drive is fear," while

$$h(y) = \ell_{hunger}[(x,y,x'),x'].$$

Applying the formula for R(d,x,t) from the previous section, we derive that

$$b(A,t) = 0$$

and

$$b(B,t) = \frac{hunger(t)}{h(y) + 1} - \frac{fear(t)}{f(t) + 1} .$$

Clearly, the animal will go to A when $b(B,t) < \varepsilon$, and will go to B when $b(B,t) > \varepsilon$, and will vacillate in between. This yields Miller's results, so long as f(y) increases more quickly than h(y) as a function of y and the curves have only one intersection point.

ACKNOWLEDGMENTS *This research was supported in part by NIH under Grant No. 5 R01 NS09755 COM of the National Institute of Neurological and Communicative Disorders and Stroke, and in part by a grant from the Alfred P. Sloan Foundation to the Center for Systems Neuroscience. Our thanks to James J. Jenkins, Richard Bogartz, and Thomas D'Andrea for drawing our attention to the relevant literature, and to Philip Gough for providing us with a copy of his 1962 paper.*

*REFERENCES*

Brown, A., Feldman, R. S., and Moore, J. W. 1968. Conditional discrimination learning based upon chlordiazepoxide. *Journal of Comparative Physiological Psychology 66:* 211-215.

Brown, J. S. 1942. The generalization of approach responses as a function of stimulus intensity and strength of motivation. *Journal of Comparative Psychology 33:* 209-226.

Brown, J. S. 1948. Gradients of approach and avoidance responses and their relation to motivation. *Journal of Comparative Psychology 41:* 450-465.

Deutsch, J. A. 1960. *The Structural Basis of Behavior*. Chicago: University of Chicago Press.

Estes, W. K. 1973. Memory and conditioning. In F. J. McGuigan and D. B. Lumsden, eds., *Contemporary Approaches to Conditioning and Learning*. New York: John Wiley & Sons.

Gough, P. B. 1962. Some tests of Hullian analysis of reasoning in the rat. *Psychonomic Science Convention* (St. Louis, 1962).

Hull, C. C. 1933. Differential habituation to internal stimuli in the albino rat. *Journal of Comparative Psychology 15:* 255-273.

Leeper, R. 1935. The role of motivation in learning: A study of the phenomenon of differential motivational control of the utilization of habit. *Journal of Genetic Psychology 46:* 3-40.

Miller, N. E. 1959. Extension of liberalized S-R theory. In S. Koch, ed., *Psychology: A Study of a Science, II*. New York: McGraw-Hill.

Strain, E. R. 1953. Establishment of an avoidance gradient under latent learning conditions. *Journal of Experimental Psyehology 46:* 391-399.

# A Model of CA1 Hippocampus as a Spatial Cognitive Map

WILLIAM L. KILMER
*Department of Computer and Information Science*
*and*
*Center for Systems Neuroscience*
*University of Massachusetts at Amherst*

DOUGLAS WILLARD*
*Center for Systems Neuroscience*
*University of Massachusetts at Amherst*

ABSTRACT *The mammalian hippocampus (HC) is posited to be an egocentric, goal-oriented, cognitive spatial mapping system. An operational scheme is developed in which, in general, the angle between an animal's direction of heading and its goal or reference object, as well as its distance from the goal, is coded by activity in an appropriate HC lamella. Some practical tests are suggested which could validate the theory.*

## 1. *INTRODUCTION*

The mammalian hippocampus (HC), shown schematically for the rabbit in Figure 1, has been posited to perform many different but related functions. Some of them are: pay attention (Stevens and Cowey, 1973; Nadel, O'Keefe, and Black, 1975) and habituate (Vinogradova, 1970); discriminate novel percepts from familiar ones (Vinogradova, 1970; Gaffan, 1972; Nadel et al., 1975); determine cue strategies (Stevens and Cowey, 1973); exert inhibitory control over attention, arousal, and movements (Bland and

---

*1975-76 Visitor, on leave from Harvard Medical School, Boston, MA.

Vanderwolf, 1972; Nadel et al., 1975); indicate matches and mismatches between actual and expected percepts (Vinogradova, 1970; Nadel et al., 1975); and select and control voluntary movements (Bland and Vanderwolf, 1972).

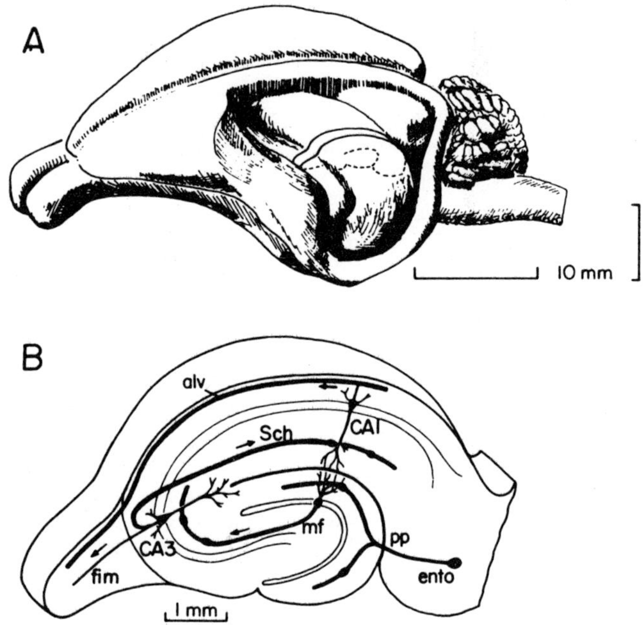


*Figure 1* Part of Figure 1 of Andersen, Bliss, and Skrede (1971).* From their caption: "A. Lateral view of the rabbit brain with the parietal and temporal neo-cortex removed to expose the hippocampal formation. The indicated section of the hippocampal formation is enlarged in B to show the main neuronal elements of the structure. Abbreviations: alv = alveus; ento = entorhinal area; fim = fimbria; mf = mossy fibres [out of dentate gyrus granule cells]; pp = perforant path; Sch = Schaffer collaterals." In a later paper (Andersen, Bland, and Dudar, 1973, Figure 10), the axonal output of the CA1 pyramidal cells is shown coursing to the right into "ento" instead of to the left into "fim." In other animals, such as rat and cat, others have indicated CA1 outputs to the fimbria/fornix system as shown above.

---

*P. Anderson, T. V. P. Bliss, and K. K. Skrede. 1971. Lamellar organization of the hippocampal excitatory pathways. *Experimental Brain Research 13:* 224. Reprinted by permission.

Ranck (1973) recorded from single HC neurons in rats free to move about in a restricted environment and found four common behavioral types of complex spike cells [presumably pyramids or dentate gyrus granule cells (cf. Figure 2)] which he characterized as follows:

> (a) *An approach-consummate cell* fired most rapidly during certain consummatory behaviors and the successful appetitive behavior associated with it. For instance, one cell fired during the rat's approach to a pellet, and while the rat explored it and then ate it. (b) *An approach-consummate-mismatch cell* fired during the same behaviors as an approach-consummate cell, but it also fired during unsuccessful behavior of the same specificity. For instance, the cell fired during the rat's approach to water and during drinking, but also during the rat's exploration of the water hole when the bottle has been removed, or when the rat was lying in front of the water hole. (c) *An appetitive cell* fired during some orienting movements or approach behavior but during no consummatory behavior except sleep. A specificity to objects or places was noted in some, but not all these cells. (d) *A motion punctuate cell* fired one to five action potentials at the end of some presumed orienting movements, or sometimes when the direction of movement was changed. Approach-consummate cells were most common in dorsal CA3. Approach-consummate-mismatch cells were most common in dorsal CA1. Appetitive cells were most common in dorsal fascia dentata. All these types were found in all regions of dorsal Ammon's horn and dorsal fascia dentata. Other less common behavioral types were also found.
>
> Most cells in presubiculum, parasubiculum, and medial entorhinal cortex are *specific orient cells* which fire for about 2 sec when the rat makes an orienting movement to something in particular, for instance food or water.*

O'Keefe (1976) also recorded from single HC neurons in freely moving rats, again in a restricted environment, and summarized

*J. B. Ranck, Jr. 1973. Studies on single neurons in dorsal hippocampal formation and septum in unrestrained rats. I. Behavioral correlates and firing repertoires. *Experimental Neurology 41:* 525-526. Reprinted by permission.

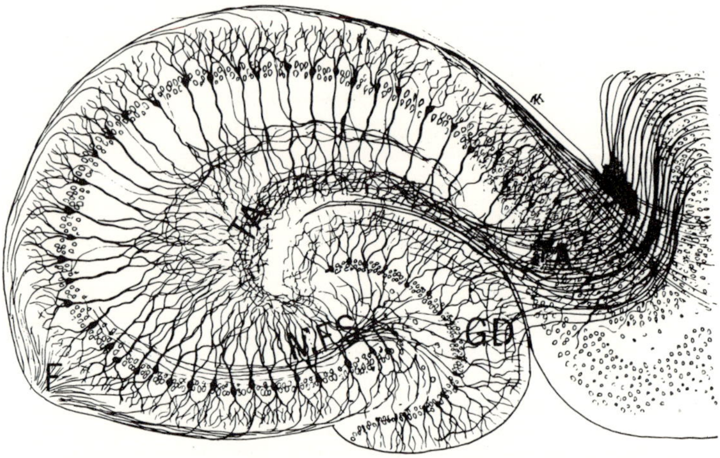


*Figure 2* A more detailed version of the lamella in Figure 1B. This picture, drawn from Golgi sections, shows granule cells of the dentate gyrus, GD, and the pyramids of the hippocampus proper, but no interneuronal cell types, in a typical transverse section of mammalian hippocampus. It is Figure 180A of Krieg (1966).* TA = temporo-ammonic tract, which originates in the generalized entorhinal region; F = fimbria/fornix; and MFS = mossy fibre system.

his findings thus:

> Single units were recorded from the CA1 field of the hippocampus in the freely-moving rat. They were classified as *place* units, *displace* units or others. Place units were defined as those for which the rat's position on the maze was a necessary condition for maximal unit firing. Some of these place units (*misplace units*) fired maximally when the animal sniffed in a place, either because it found something new there or failed to find something which was usually there. Displace units increased their rates during behaviors associated with theta activity in the hippocampal slow waves. In general these were behaviors

---

*W. J. S. Krieg. 1966. *Functional Neuroanatomy,* 3rd ed. Evanston, Illinois: Brain Books, p. 348. Reprinted by permission.

> which changed the rat's position relative to the environment. The influence of various environmental manipulations (e.g., turning off the room lights) on the firing pattern of the place units was tested and the results suggest that they were not responding to a simple sensory stimulus nor to a specific motor behavior. Nor could the unit firing be due purely to motivational or incentive factors. The results are interpreted as strong support for the cognitive map theory of hippocampal function.*

From the work surrounding the foregoing positions and findings, it seems clear that HC is mainly concerned with exteroceptive rather than interoceptive conditions; is not a repository of reflex pathways; is not bound to sensory stimuli or specific animal movements to the degree found, for example, in motor and sensory cortices; and is concerned with the formation of memory-based linkages between perceptual constellations and goal-oriented actions. It therefore must attend, habituate, remember, and direct overt behavior of the animal with respect to goal locations in its living space. HC thus probably qualifies as a cognitive computer, at least in its final CA1 stage. The large input to CA1 from various cortical regions, as elucidated anatomically by Van Hoesen, Pandya, and Butters (1972) (see Figure 3), corroborates this conclusion.

### 2. *CA1 HC AS A COGNITIVE SPATIAL MAPPING SYSTEM*

Figure 4 summarizes an experiment which contributed to O'Keefe's early hypothesis (O'Keefe and Dostrovsky, 1971) that CA1 HC is part of the brain's embodiment of an egocentric spatial cognitive map. It shows an HC cell that responded vigorously only when a struggling rat assumed a certain angular heading with respect to its escape route. The general idea of an animal keeping close track of the angle between its heading and its goal has been corroborated by over five years of daily behavioral experimentation

---

*J. O'Keefe. 1976. Place units in the hippocampus of the freely-moving rat. *Experimental Neurology 51:* 78. Reprinted by permission.

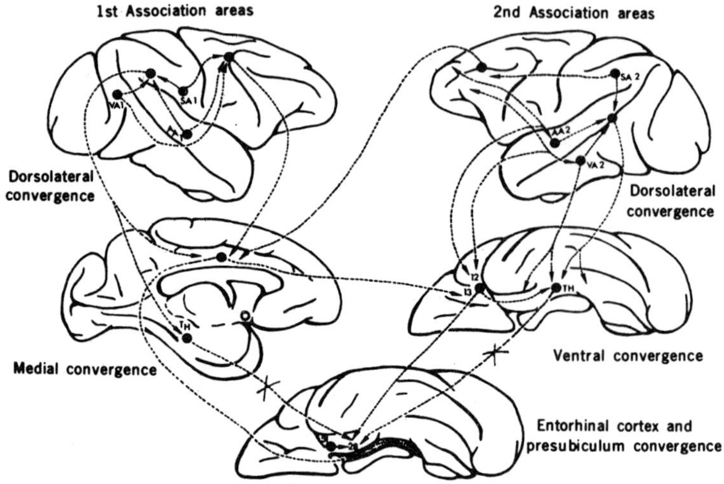


*Figure 3* Figure 2 of Van Hoesen, Pandya, and Butters (1972)* on HC anatomical inputs in the rhesus monkey. Their caption: "The probable multisynaptic pathways via which the hippocampus of the limbic system has access to sensory information. These pathways are depicted in the dorso-lateral, medial, and ventral views of the rhesus monkey brain. Sites of origin (circles) and termination (arrows) refer only to general cytoarchitectonic regions and not to specific points. Primary association areas are labeled *VA*1, the area for vision; *SA*1, that for somesthesis; and *AA*1, that for audition. Secondary association areas are labeled similarly, except that the digit is changed to 2." TH is temporal cortex, adjacent to medial entorhinal cortex, area 28. Areas 12 and 13 are orbital frontal neocortex, and area 51 is prepiriform cortex.

with horses (Kilmer, in preparation). This idea, if generally correct, also provides a coherent explanation for the seemingly disparate theories of HC function mentioned above.

Kilmer and McLardy (1971) posited that one function of the temporo-ammonic input to HC (cf. Figure 2) is to relatively boost and depress the excitabilities of different HC subcircuits. These subcircuits were associated with gross modes and specific aims of animal behavior. We now posit that CA1 HC has facilitatively

---

*G. W. Van Hoesen, D. N. Pandya, and N. Butters. 1972. Cortical afferents to the entorhinal cortex of the rhesus monkey. *Science* *175:* 1472. Copyright 1972 by the American Association for the Advancement of Science. Reprinted by permission.

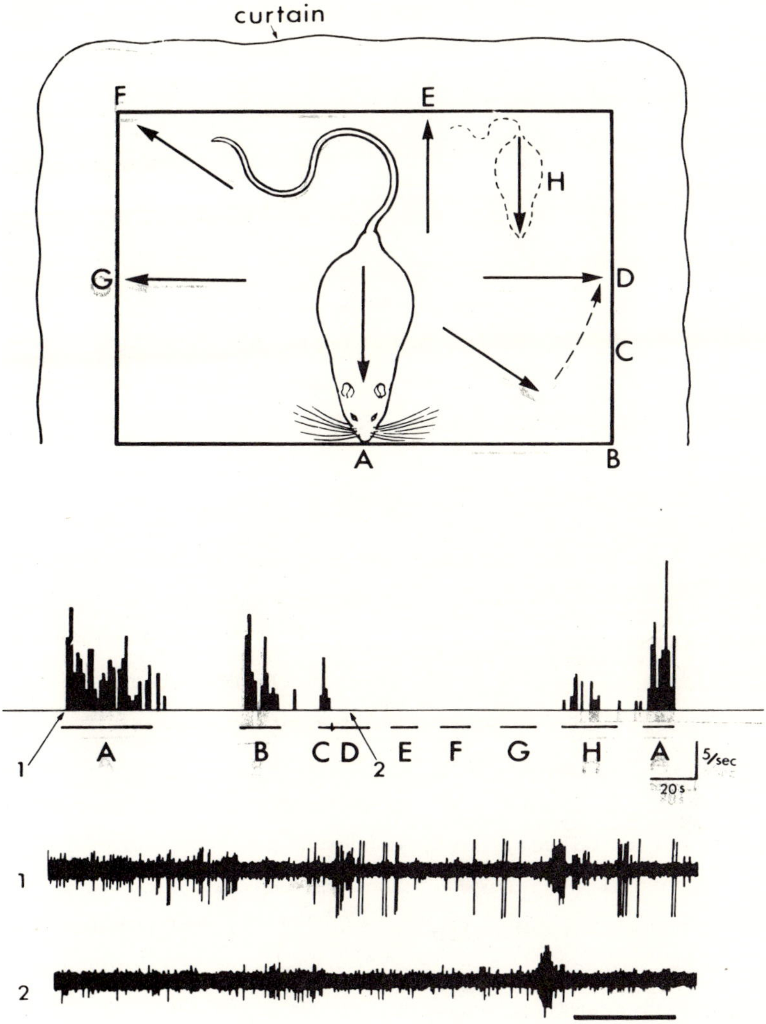


*Figure 4* Figure 1 of O'Keefe and Dostrovsky (1971).* Their caption: "Responses of a hippocampal (CA1) unit to a restraining tactile stimulus as a function of the rat's spatial orientation. The arrows and associated letters mark the positions at which the animal was restrained as it was pushed or coaxed in a counterclockwise direction around the test platform. The firing rate of

---

*J. O'Keefe and J. Dostrovsky. 1971. The hippocampus as a spatial map. Preliminary evidence from unit activity in the freely-moving rat. *Brain Research 34:* 173. Reprinted by permission.

linked maps for food, water, home, hiding places, etc. This complements Kilmer and McLardy's (1971) and Kilmer and Olinsky's (1974) previous notion that CA3 HC is a more perceptual-motor decisionary computer (of a type that is diffusely preprogrammed at birth and filled in by experiences during its behavior-formative period). This CA3 concept involved less temporal structure in its computations than is now posited for CA1. With this general framework in mind, we shall attempt to specify how the spatial mapping functions of CA1 HC might be pushed down into its circuit organization without violating any data currently available, and thence to specify what sorts of experiments ought to be performed next.

## 3. *CONCEPTUAL FOCUS*

We shall focus on the central notions of the cognitive mapping problem with two especially appropriate animals for studying it in mind: the rabbit and the horse. The solitary aspects of these animals' spatial behaviors involve little more than eating, drinking, escaping from predators, and homing (to a nest or seasonal territory). Neither hunts prey, so neither possesses the tracking, shortcutting, ambushing, and deception strategies to the degree found in wolves, cats, baboons, and other predators. Horses and rabbits do have escape strategies, such as crisscrossing in front of pursuers (rabbits), and erasing trails over midriver paths (horses). But usually they strive to avoid predators through evasion and concealment, which might involve freezing or running to hiding places (rabbits), or running away for a mile or

---

the unit during this procedure is illustrated by the continuous frequency histogram in the middle of the figure. The letters correspond to the positions and the lines indicate the periods when the rat was restrained. In between these periods, the rat sat immobile in the same position for a few seconds and then was moved on to the next position. The bottom two lines show the rat data taken at the onset of the unit response at A(1) and during the absence of a response at D(2). Time calibration for these data is 400 msec."

so (horses). Both animals especially avoid those *places* where major accidents or threats (real or imagined) have occurred before. Also, the scales and qualities of their living spaces are better matched to our own cognitive mapping abilities than is the case for rats (especially the climbing type), moles, dolphins, or tree shrews, for example. The HC's of both animals constitute relatively large fractions of their brains. Kilmer and McLardy, in a pilot neuroanatomical investigation, found the horse to have a CA1 HC that is lengthy but shallow in pyramidal cell depth, suggesting a paucity of goal/aim categories. Finally, much is known about these animals' drives and associated locomotor behaviors.

Figure 5 shows an example of a homing path for a rabbit or horse. We shall first assume that our animal has learned the path in Figure 5 very well from past experience, and then later consider how this could have been brought about. As the animal travels from 1' to 4, we note that it apparently performs three different types of movement with respect to the distance from X and the angle between the animal's heading and X. The caption to Figure 6, however, clearly shows that all three movements are subject to the same $\Delta\alpha$ and $\Delta D$ formulae. These formulae involve only distance and angle to the goal, speed of locomotion, and sine and cosine of the angle to the goal.

The curved movement from 4 to 4' (Figure 5) can be regarded as a succession of many short transitions of the 3-to-3' type. The special end-path traversal from 4' to X, and the different opening-path traversal previous to 1', are discussed in Section 5.

The aim of the next section will be to abstract a spatial cognitive mapping model for CA1 HC that could represent the guiding agent of our animal along the path in Figure 5. We shall assume a general familiarity with HC neuroanatomy and neurophysiology and suppose that the requisite pathways and information flows exist, as indeed they appear to.

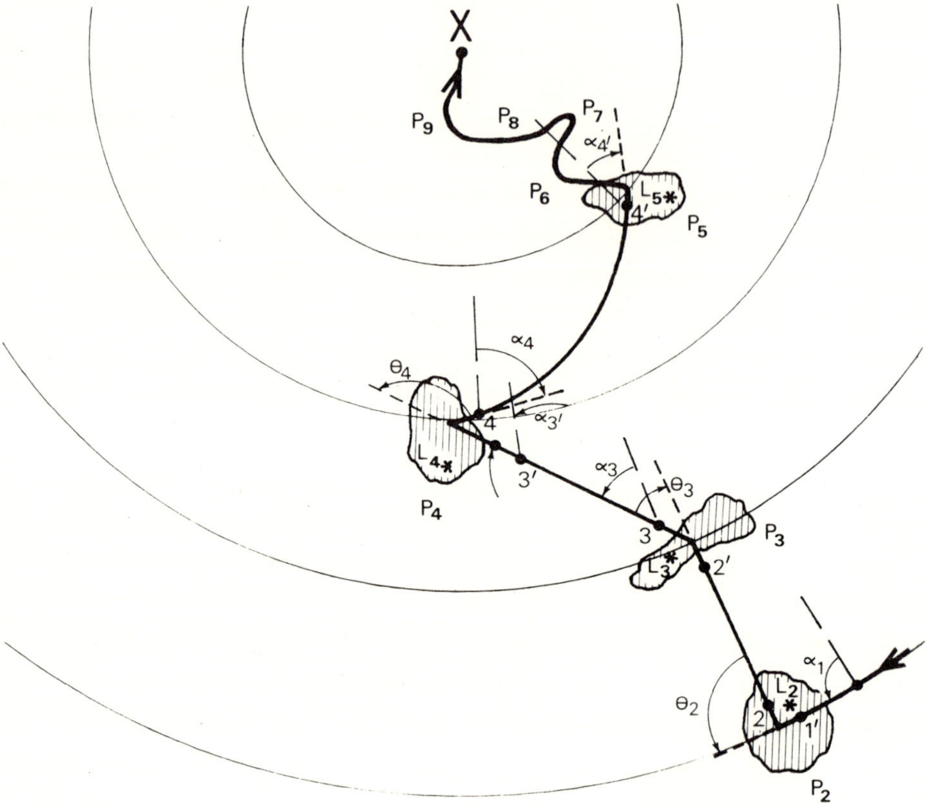


*Figure 5* A hypothetical, well-learned path taken by a rabbit or horse with respect to a goal, such as water or "home," at X. The path from 1' (lower right) to X might be a small city block length for a rabbit, or a mile or two for a horse. Each $\alpha_i$ is the angle between the animal's heading and X, and each $\theta_i$ is the angle by which the animal's direction is changed before the path point i. $P_i$ indicates a multimodal perceptual "scene," and $L_i$ a focal landmark in that scene. The map is assumed drawn to scale.

## 4. *THE BASIC CA1 MODEL*

Our model assumes that the mammalian hippocampal CA1 acts inter-alia as an egocentric spatial cognitive map. Figure 7 shows our schematization for an animal's entire bilateral CA1 hippocampus. The annulus is not broken at the ventrolateral extremity, even though the right and left ventral-lateral hippocampi do not touch, because they are sufficiently connected through the hippocampal commissures to permit them to be effectively continuous (Hjorth-Simonsen and Zimmer, 1975). Each sector of the annulus corresponds to the CA1 portion of a hippocampal lamella as shown in the lower right of the figure. The right/leftness of the representations in the model can be reversed without any essential change in the theory.

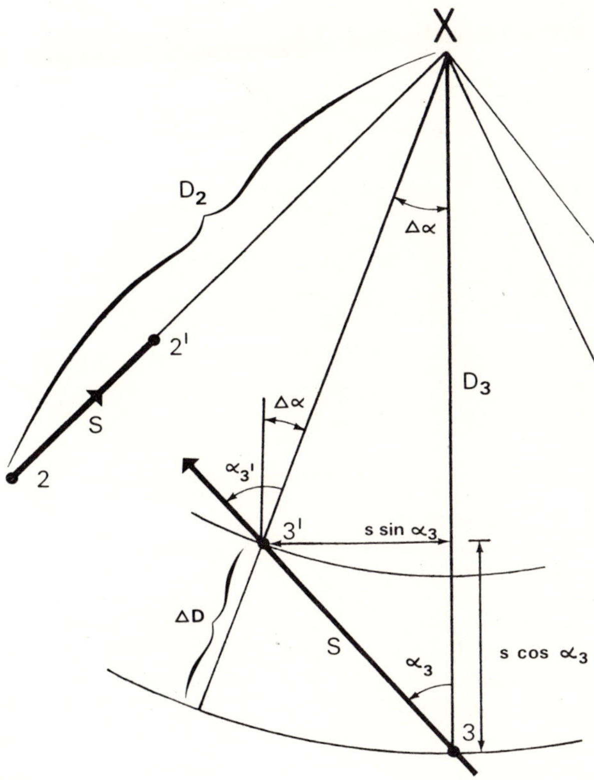


*Figure 6* Elements of the three kinds of movement in Figure 5 along the path from point 1' to 4, at least as far as the $\alpha_i$ and distances $D_i$ from points i to X are concerned. The positive senses of $\alpha_i$ and $\theta_i$ are counterclockwise. The general case is the movement from 3 to 3', where $\Delta\alpha$ in nadian units is approximately $(s \sin \alpha_3)/D_3 = C(s,\alpha_3,D_3)$, and $\Delta D_3 \doteq s \cos \alpha_3 = C_D(s,\alpha_3)$. These formulae are correct for all $\alpha_3$ between $-180^\circ$ and $+180^\circ$.

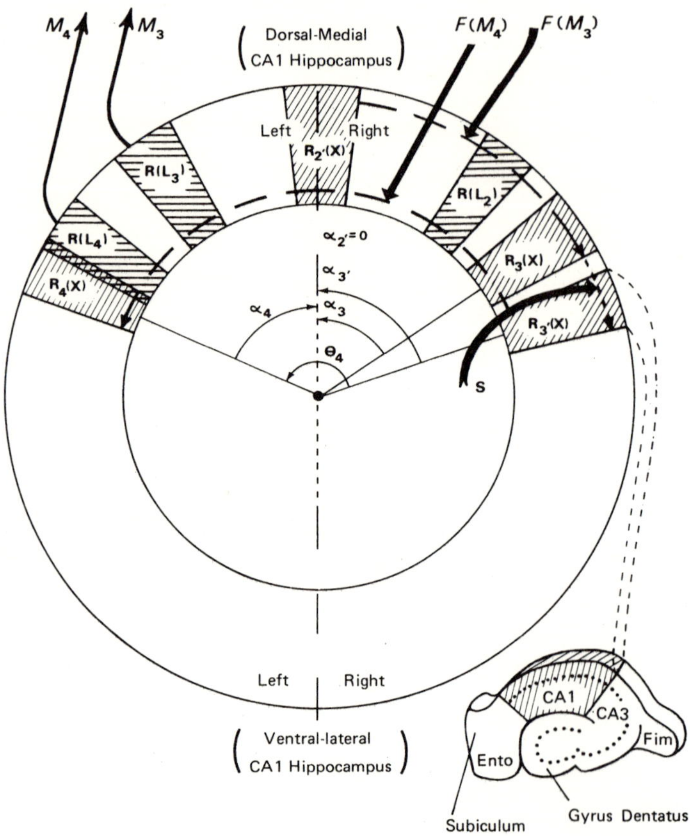


*Figure 7* Schematic of the CA1 model. See text for explanation. The lamella at lower right is the left-to-right reverse of the lamellae in Figures 1 and 4 since it belongs in the right HC and not the left.

We assume that there are sufficient ipsilateral longitudinal connections within CA1 so that all posited circumferential translations of circuit activity within the annular model are plausible. We also assume that neural representation of complex percepts centered upon landmark objects $L_i$ in extrapersonal space are somatotopically transmitted from entorhinal cortex, subiculum, and CA3 into CA1 (Andersen, Bliss, and Skrede, 1971; Van Hoesen et al., 1972). By "somatotopically" we mean "with respect to the difference between the direction the animal is facing and the direction of its goal X." We further assume that each lamellar slice of CA1 can, without confusion, support activities representing: X [denoted R(X)]; the approximate distance, D, the animal now is from X; the actual perception of an $L_i$ [denoted $R(L_i)$]; the prediction of (or search for) a perception of an $L_i$; and the coherent movement of any representation R(.) through the lamella by a feedback signal from the animal's turning circuitry [denoted F(M)] or stepping circuitry. The $M_i$ represents signals from the $R(L_i)$ regions to the animal's motor circuits (in the basal forebrain and brainstem) (Siegel and Tassoni, 1971) that serve to turn the animal's heading with respect to X through the angle $\theta_i$.

We shall explain the operation of the model by tracing its representation of the animal's passage in Figure 5 along a well-learned path from point 2', through points 3 and 3', to point 4. We assume that along this path neither is X sensible (visible, audible, or "smellible"), nor are $L_3$ and $L_4$ close enough to be simultaneously sensible. We assume, though, that $D_{2'}$, the path distance from 2' to X, is small enough so that $(\alpha_3/\alpha_{3'})$ and $(D_4/D_{2'})$ are fractions appreciably less than 1.0.

We start with the animal at 2' and heading directly toward home, our goal X in this case. We posit that X's direction relative to the direction the animal is headed is represented by the angle in the annular model from R(X) to the upward vertical originating at the center of the annulus. Thus at 2', $R_{2'}(X)$ is located as shown in Figure 7.

The animal next perceives the landmark object $L_3$ to the left of 2', as shown in Figure 5. The corresponding representation in the model, denoted $R(L_3)$, is shown to the left of $R_{2'}(X)$ in Figure 7. Presumably, $R(L_3)$ is infixed partly over the neocortical → temporo-ammonic pathway (Van Hoesen et al., 1972), and partly by predictor signals into the $R(L_3)$ region from the $R_{2'}(X)$ and $R(L_2)$ regions. Thus, the animal anticipates the appearance of

$L_3$ by circuit actions which are suggestive of Ranck's (1973) approach cells and O'Keefe's (1974) place units. X's for O'Keefe might have been extramaze reference points instead of goals in the usual sense, and angles-to-X's for O'Keefe might sometimes have been the angle between X and the direction his rat would have had to take to get to X along either a beeline or some other path. Returning to our model, if the animal fails to find $L_3$ as expected, it generates an anticipatory mismatch signal pattern in the $R(L_3)$ region of the type found by Ranck (in his mismatch cells) or by O'Keefe (in his misplace cells).

$R(L_3)$ next transmits the previously learned signal $M_3$ to the animal's motor circuits to cause the animal to change direction by turning through the angle $\theta_3$. Doing so gives rise to the feedback signal $F(M_3)$, which translates the $R_{2'}(X)$ activity circumferentially along the model to the $R_3(X)$ position located at the angle $\alpha_3$ from the vertical. Presumably, $F(M_3)$ is generated in the basal ganglia or midbrain and is transmitted to CA1 via the septum or directly over the fornix (Ibata, Desiraju, and Pappas, 1971; Siegel and Tassoni, 1971). Perhaps O'Keefe's displace cells help to translate $R_i(X)$ activities axially in HC. The situation at 3 in Figure 5, then, is represented in Figure 7 by $R_3(X)$, the fresh memory trace $R(L_3)$, and a moderately accurate representation $R(D_3)$ of the path distance from 3 to X.

The path in Figure 5 from 3 to 3' is of the second type in Figure 6. Thus, as the animal steps along from 3 to 3', a cerebellar (Heath and Harper, 1974) or midbrain motor input to CA1 (Siegel and Tassoni, 1971), represented by S in Figure 7, translates $R_3(X)$ to $R_{3'}(X)$ in successive $\Delta$ increments according to the $\Delta\alpha$ function $C_2$ given in Figure 6. We posit that the $\alpha_3$ constraint on $R_3(X)$, an $R_3(X)$-lamellar activity $R(D_3)$, and the S input to $R_3(X)$ interact to realize the $C_\alpha$ and $C_D$ functions of Figure 3 at $R_3(X)$ in Figure 7. Many repetitions of this interaction yields $R_{3'}(X)$ as shown in Figure 7, with an appropriate associated $R(D_{3'})$. If the distance 3 to 3' were, say, one third $D_3$, $(\alpha_3/\alpha_{3'})$ and $(D_{3'}/D_3)$ would both be appreciably less than 1. Finally, $R(L_3)$ and $R_{3'}(X)$ predict $R(L_4)$, which upon being found gives rise to $M_4$, whence follows $F(M_4)$ and the translation of $R_{3'}(X)$ to $R_4(X)$.

## 5. *OTHER FEATURES OF THE MODEL*

In the previous section, we assumed that $D_2'$ was neither so large as to be effectively infinite in a low-precision $\Delta\alpha$ computation, nor so small that the animal could perceive one $L_i$, together with its associated perceptual context, before losing every facet of the direct perception of $L_{i-1}$. In Figure 5, the path prior to 1' is assumed to be of the large D type, and the path from 4' to X is short enough for the animal to be able to rely on only exteroceptive and proprioceptive cues to guide it to X.

Navigation at large D is probably facilitated in almost all animals by their following well-worn or scented paths between widely spaced $L_i$, or else by their traveling from lookout point to lookout point over an easily remembered terrain pattern, such as a hill-valley-hill sequence. In the latter case, they might also be guided by the sun, wind direction, vegetation pattern, etc. In terms of our Figure-7 model, navigation is implemented mostly by rather vague $R(L_i)$ chains. This emphasizes that the $R(L_i)$ can range from very sharp representations of very striking objects to rather amorphous representations of large multimodal "scenes." Such scenes would have spatial aspects themselves, and these might be somewhat somatotopically represented within single lamellae.

We assume that an animal can bridge from one Figure-7 goal map to another (from X = home to X = water, say) through the relative depression of one CA1 subnet (perhaps via CA1 inputs over the temporo-ammonic tract) and the concomitant excitation of another. The active $P(L_i)$ regions would remain active throughout the transition, and R(new X) would be generated by R(old X) and the $P(L_i)$, there being, in general, several of the latter.

Note that the $M_i$ outputs in Figure 7 must be able to code any value of $\theta_i$. Perhaps this, in part, explains the very large size of the fimbria-fornix (Raisman, Cowan, and Powell, 1965; Brown and Winocur, 1972).

A Figure-7 map could be learned in reverse order to the direction of its Figure-5 path by an animal's repeatedly exploring the area further and further out from X. In doing this, the animal would have to go confer goal status on representations of lengthening lists of $L_i$ (cf. Arbib and Lieblich, this volume). It might also, in the process, store representations of D as inherent activity levels. The cingulate cortex would thus be enabled to learn a telescoped version of Figure 5 by remembering only the sequence of successive $[P(L_i),\theta_i]$ points. From this, shortcuts could be learned in higher cortex by collapsing sequences like

$$[P(L_i),\theta_i],\ [P(L_{i+1}),\theta_{i+1}],\ \ldots,\ [P(L_{i+3}),\theta_{i+3}]$$

to

$$[P(L_i),\theta_i],\ [P(L_{i+3}),\theta_{i+3}]$$

(perhaps rendering CA1 to some extent redundant in wildlife adulthood, as suggested some time ago by McLardy, 1971). Thus, the model suggests a proper "mental geometry" instead of the Euclidean geometry used in most current artificial intelligence work.

## 6. *EXPERIMENTAL TESTS*

The afferent, efferent, and intrinsic neuroanatomy of CA1 needs further elucidation, especially the ipsilateral longitudinal component, if our model is to be validated. Also, since we see best ahead of us, hear best to the side of us, and smell best at close range all about us, the dorsal-ventral distribution in CA1 over these modalities might, if elucidated, lend strong indirect support to our Figure-7 scheme. Most pressing of all, perhaps, is the need to record from a battery of electrodes along the different lamellar slices of CA1 in animals free to move about. Ablation of the hippocampus on only one side or a hippocampal commissurotomy should furthermore lead to serious spatial confusion whenever the animal turned enough for its goal or reference point X to swing from one side of it to the other side. The left/right sidedness of the model could be explored in this way, too. The possibility that O'Keefe's place cells have more to do with the angle between X and the direction an animal would have to take to get to X, along either a beeline or some other path, should also be checked, as well as the possibility that X's might change quite frequently in some animals. Rabbits in an essentially natural environment would be excellent subjects for such experiments, and so would domestic horses, about which an enormous amount of behavioral information is available.

If our model is correct, the current testing technique of rotating rats on maze platforms with respect to their extramaze environments confuses the issue because most of an animal's spatial cognitive operations are based on internal representations of how its previous locomotor movements have affected its record of where it is with respect to its present goal or reference point. Also, moving goal objects about in restricted environments and requiring animals to solve local extrapersonal spatial problems do not really test the ideas in this paper, but instead test object tracking, primitive planning, and local reorienting abilities.

Single HC cell records taken from preferably immature animals in the process of acquiring their spatial maps should be an invaluable guide to a more fine-grained formulation of our $R(L_i)$ landmark representations for the tested species.

*REFERENCES*

Andersen, P., Bland, B. H., and Dudar, J. D. 1973. Organization of the hippocampal output. *Experimental Brain Research 17:* 152-168.

Andersen, P., Bliss, T. V. P., and Skrede, K. K. 1971. Lamellar organization of hippocampal excitatory pathways. *Experimental Brain Research 13:* 222-238.

Bland, B. H., and Vanderwolf, C. H. 1972. Diecephalic and hippocampal mechanisms of motor activity in the rat: Effects of posterior hypothalamic stimulation on behavior and hippocampal slow wave activity. *Brain Research 43:* 67-88.

Brown, R. J., and Winocur, G. 1972. The fornix as a reward pathway. *Physiology and Behavior 11:* 47-52.

Gaffan, D. 1972. Loss of recognition memory in rats with lesions of the fornix. *Neuropsychologia 10:* 327-341.

Heath, R., and Harper, J. 1974. Ascending projections of the cerebellar fastigial nucleus to the hippocampus, amygdala, and other temporal lobe sites: Evoked potential and histological studies in monkeys and cats. *Experimental Neurology 45:* 268-287.

Hjorth-Simonsen, A., and Zimmer, J. 1975. Crossed pathways from entorhinal area to the fascia dentata in normal rabbits. *Journal of Comparative Neurology 161:* 57-70.

Ibata, Y., Desiraju, T., and Pappas, G. D. 1971. Light and electron microscopic study of the projection of the medial septal nucleus to the hippocampus of the cat. *Experimental Neurology 33:* 103-117.

Kilmer, W. L., and McLardy, T. 1971. A diffusely preprogrammed but sharply trainable hippocampus model. *International Journal of Neuroscience 2:* 241-248.

Kilmer, W. L., and Olinski, M. 1974. Model of a plausible learning scheme for CA3 hippocampus. *Kybernetik 16:* 133-144.

Krieg, W. J. S. 1966. *Functional Neuroanatomy,* 3rd ed. Evanston, Illinois: Brain Books.

McLardy, T. 1971. Anticipatory recall deficit after cingulumotomy in rats. *Experimental Neurology 32:* 141-151.

Nadel, L., O'Keefe, J., and Black, A. 1975. Slam on the brakes: A critique of Altman, Brunner, and Bayer's response-inhibition model of hippocampal function. *Behavioral Biology 14:* 100-107.

O'Keefe, J. 1976. Place units in the hippocampus of the freely moving rat. *Experimental Neurology 51:* 78-109.

O'Keefe, J., and Dostrovsky, J. 1971. The hippocampus as a spatial map. Preliminary evidence from unit activity in the freely-moving rat. *Brain Research 34:* 171-175.

Raisman, G., Cowan, W. M., and Powell, T. P. S. 1965. The extrinsic afferent, commisural and association fibres of the hippocampus. *Brain 88:* 963-995.

Ranck, J. B., Jr. 1973. Studies on single neurons in dorsal hippocampal formation and septum in unrestrained rats. I. Behavioral correlates and firing repertoires. *Experimental Neurology 41:* 461-555.

Siegel, A., and Tassoni, J. 1971. Differential efferent projections from the ventral and dorsal hippocampus of the cat. *Brain, Behavior and Evolution 4:* 185-200.

Stevens, R., and Cowey, A. 1973. Effects of dorsal and ventral hippocampal lesions on spontaneous alternation, learned alternation and probability learning in rats. *Brain Research 52:* 203-224.

Van Hoesen, G. W., Pandya, D. N., and Butters, N. 1972. Cortical afferents to the entorhinal cortex of the rhesus monkey. *Science 175:* 1471-1473.

Vinogradova, O. S. 1970. Registration of information and the limbic system. In G. Horn and R. A. Hinde, eds., *Short-term Changes in Neural Activity and Behavior*. Cambridge: Cambridge University Press.

# Author Index

Numbers in italics refer to the pages on which the complete references appear.

## D



## E



## F



## G



## H



## I



## J

## K



## L



## M



## N



## O



## P

# Subject Index

## A



## B



## C

## D



## E



## F



## G



## H

## O



## P



## R

## S



## T

## V



## W



## X



## Y



A
B 7
C 8
D 9
E 0
F 1
G 2
H 3
I 4
J 5